QUANTUM THEORY OF CHEMICAL REACTIONS

# QUANTUM THEORY OF CHEMICAL REACTIONS

## *III. Chemisorption, Catalysis, Biochemical Reactions*


*Edited by*

RAYMOND DAUDEL
*CNRS, Centre de Mécanique Ondulatoire Appliquée, Paris, France*

and

ALBERTE PULLMAN
*CNRS, Institut de Biologie Physico-Chimique, Paris, France*

LIONEL SALEM
*CNRS, Laboratoire de Chimie Théorique, Université de Paris Sud, Orsay, France*

ALAIN VEILLARD
*CNRS, Université Louis Pasteur, Strasbourg, France*



D. REIDEL PUBLISHING COMPANY
DORDRECHT : HOLLAND / BOSTON : U.S.A.
LONDON / ENGLAND

**Library of Congress Cataloging in Publication Data** (Revised)
Main entry under title:

Quantum theory of chemical reactions.

Includes bibliographical references and index.
CONTENTS: v. 1. Collision theory, reaction path, static indices.–v. 2 Solvent effect, reaction mechanisms, photochemical processes–v. 3. Chemisorption, catalysis, biochemical reactions.
1. Quantum chemistry–Addresses, essays, lectures. I. Daudel, Raymond.
QD462.5.Q38 541.2′8 79-22914
ISBN 90-277-1047-3 (v. 1)
ISBN 90-277-1182-8 (v. 2)
ISBN 90-277-1467-3 (v. 3)

---

Published by D. Reidel Publishing Company,
P.O. Box 17, 3300 AA Dordrecht, Holland.

Sold and distributed in the U.S.A. and Canada
by Kluwer Boston Inc.,
190 Old Derby Street, Hingham, MA 02043, U.S.A.

In all other countries, sold and distributed
by Kluwer Academic Publishers Group,
P.O. Box 322, 3300 AH Dordrecht, Holland.

D. Reidel Publishing Company is a member of the Kluwer Group.



Printed in The Netherlands

## TABLE OF CONTENTS

# PREFACE

The third and last volume of this treatise is concerned with important applications of the quantum theory of chemical reactions to chemisorption, catalysis and biochemical reactions.

The book begins with an important paper devoted to the theoretical background of heterogeneous catalysis. It is followed by two papers showing typical applications of wave mechanics to the analysis of chemisorption. Catalysed gas-solid reactions are chosen to illustrate gas, organic solid state reaction and some aspects of the mechanism of the FISCHER-TROPSCH synthesis are presented.

The second part of the book is devoted to biochemical applications of quantum chemistry. Two papers are concerned with the quantum theory of enzyme activity. Two others present recent progress of quantum pharmacology.

Finally an important contribution to the theory of intermolecular forces is made in the view of possible applications to biochemical problems.



*R. Daudel, A. Pullman, L. Salem, and A. Viellard (eds.), Quantum Theory of Chemical Reactions, Volume III, vii.*

# THEORETICAL BACKGROUND OF HETEROGENEOUS CATALYSIS

J.E.Germain
Laboratoire de Catalyse Appliquée et Cinétique Hétérogène
L.A. 231 du Centre National de la Recherche Scientifique
Université Claude Bernard Lyon I, E.S.C.I.L.
43 Boulevard du 11 Novembre 1918, 69622 Villeurbanne Cédex.

Heterogeneous Catalysis is a surface Kinetic phenomenon by which a chemical reaction between molecules of a fluid phase is accelerated (activity) and oriented (selectivity) by contact with a solid phase (catalysts, without change of the solid.

We shall mostly discuss the case of metal catalysts.

## I. FUNDAMENTAL PROCESSES.

Perfectly described by I.Langmuir in the years 1920-30, they include : adsorption, surface reaction, desörption.

### a) Chemical adsorption (or Chemisorption).

The clean surface of a metal traps selectively gas molecules with a large heat release (20-100 kcal/mole) around 300K. The number of molecules thus adsorbed per unit surface area is limited (saturation) ; in simple cases (CO, $H_2$, $O_2$), stoechiometric relationships are found between this number and the atomic density of usual crystal faces ($N \;\; 10^{15}/cm^2$).

There facts are best explained by chemical reaction of these molecules with unsaturated adsorption sites created by disrüption of the threedimensional crystal lattice ("free valencies"). By analogy with classical chemistry, one predicts that unsaturated molecules (CO, $C_2H_4$) may adsorb directly (associative adsorption) while saturated molecules ($H_2$, $CH_4$) must break into radicals (dissociative adsorption). This is generaly the case.

The following energy scheme shows that several adsorption states (1,2,3...) are available ; at a given temperature, the only accessible states are the ones such that activation barriers ($E^x$) are not too high ($E^x/RT$ factor) (Fig. 1).

The favored equilibrium state will be the lowest of these accessible states (several such states of similar energies could exist).



*R. Daudel, A. Pullman, L. Salem, and A. Viellard (eds.), Quantum Theory of Chemical Reactions, Volume III, 1–13.*

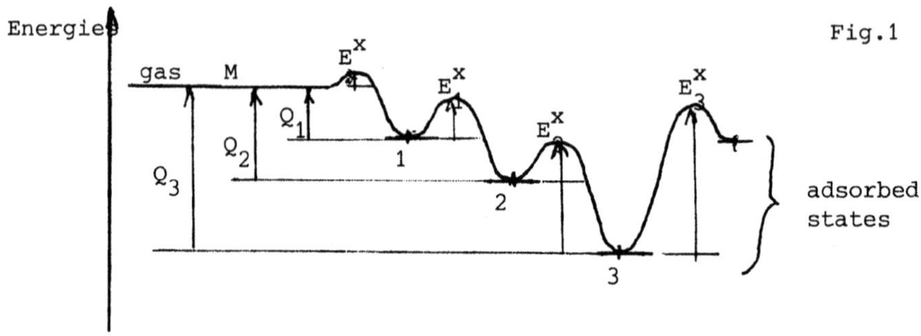

Fig.1

Remarks : - the first accessible state usualy corresponds to associative adsorption.
- on metal surfaces, the first activation energic ($E^x$) is frequently very low.
- the transition between 2 states is an elementary step in the sense of kinetics.

The following examples illustrate particular cases (Fig. 2).

| | | |
|---|---|---|
| CO/Ni | 300 K | non activated associative adsorption |
| $H_2$/Pt | 300 K | non activated dissociative adsorption |
| $N_2$/Fe | 500 K | activated dissociative adsorption |

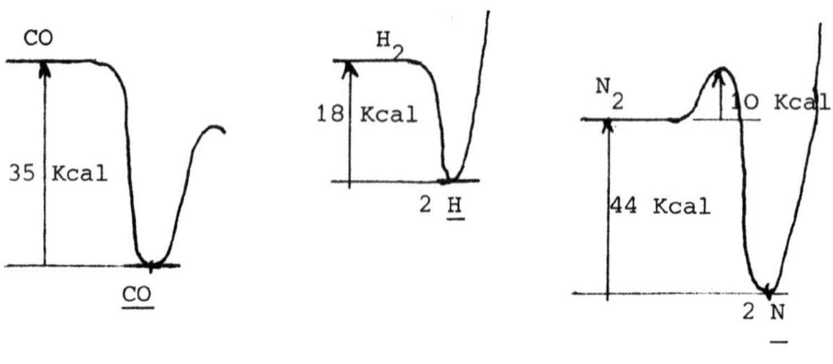

Fig.2

Dissociative adsorption may lead to an adsorbed species plus a gaseous species :

$NH_3$/Fe → $\underline{N}$ + $H_2$
$N_2O$/Cu → $\underline{O}$ + $N_2$

As predicted by the general scheme (Fig. 1), the state of adsorption may change when temperature is raised as the number of accessible states increase.
e.g. (Fig. 3) : $NH_3$/Fe

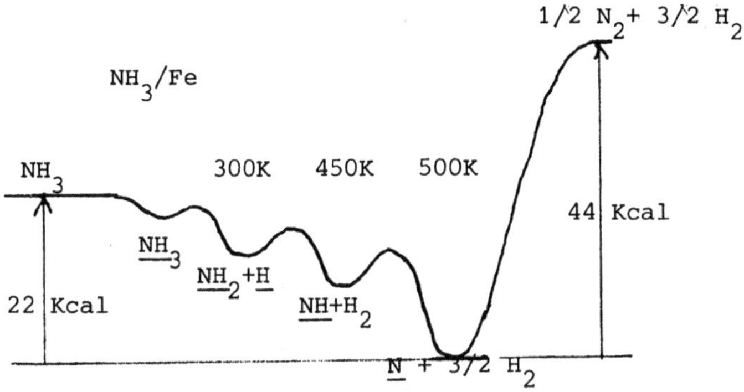


Fig.3

Prediction of these energy schemes (potential surfaces) for a given gas-solid system is not yet possible ; the specific states of each catalyst is given by experience.

We shall point out the remarkable activity of transition metals, capable to dissociate, at room temperature, very stable molecules ($H_2$, $O_2$, $CH_4$,...) and the high binding energies between atoms and sites ($q_{H-M}$ = 60 Kcal/M on Ni).

At higher temperatures, adsorbed atoms may dissolve in the metal (C, N in α-Fe, O in Ni) or produce defined chemical compounds (carbide, oxide).

b) Catalysis.

According to thermodynamics, the desorption (endothermal) is possible at sufficiently high temperature, and suffcicently low pressure ($P < 1$ Torr).

Thus, if Q = 20 Kcal/M at T = 400-500 K
Q = 100 Kcal/M at T = 2000-2500 K

But at these temperatures new gas phase states become also accessible for the system, and the initial adsorbed gas molecules are not necesarily restituted : desorption is not univocal.

ex. (Fig. 4) $H_2$/Pt "reversible" adsorption
$NO$/Pt "irreversible" adsorption

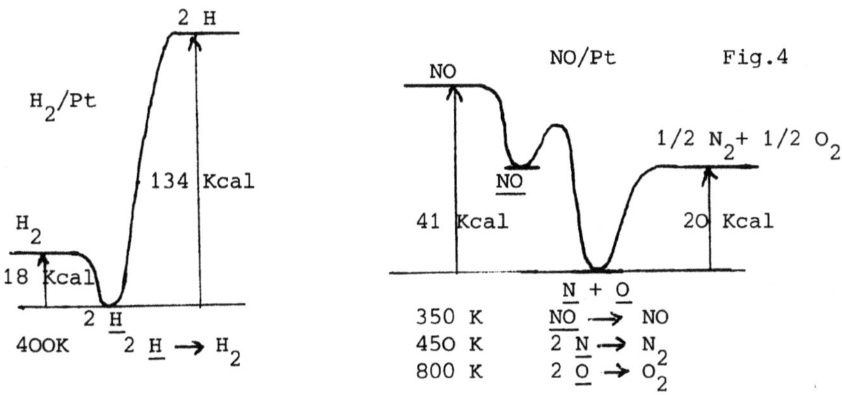


Fig.4

The possibility of decomposition (or isomerization) catalysis arises. We may replace the cycle : low temperature adsorption of M, high temperature desorption, by the contact of M with the solid at a temperature T such that desorption is possible in 2 directions. A continuous decomposition (Fig. 5)

$$M \rightarrow P + Q$$

will be observed, by maintaining a constant pressure of M, since P and Q with desorb freely.

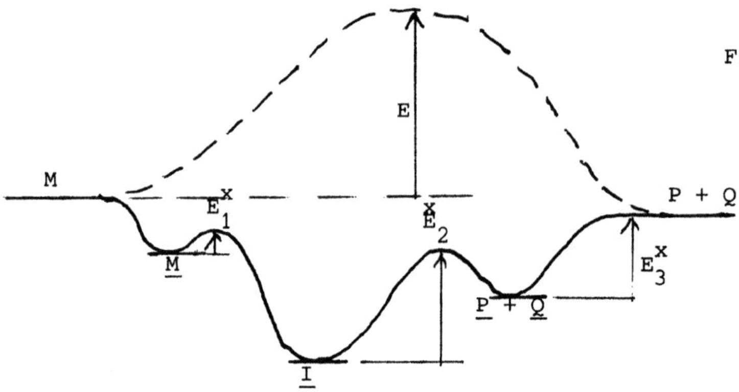


Fig.5

This reaction will be accelerated if the activation barrier E of the homogeneous reaction is replaced by a series of lower barriers ($E_1^x$, $E_2^x$, $E_3^x$).

The reaction is also oriented since P and Q depend of the nature of the adsorbed intermediate I, characteristic of the catalyst.

The above scheme has been simplified by grouping the formation of products P and Q, which are in fact distinct processes.

ex. $\underset{(M)}{NO} \rightarrow \underset{(P)}{N_2} + \underset{(Q)}{\frac{1}{2} O_2}$

$$\underset{(N)}{NO} \rightarrow \underset{(M)}{\underline{NO}} \rightarrow \underset{(I)}{\underline{N} + \underline{O}} \quad \begin{cases} 2\,\underline{N} \rightarrow N_2 & (P) \\ 2\,\underline{O} \rightarrow O_2 & (Q) \end{cases}$$

The principle of microscopic reversibility indicates that by changing the direction of each elementary step, we obtain the mechanism of catalytic synthesis :

$$P + Q \rightarrow M$$

In every case, we may distinguish the following steps :

$$\begin{cases} \text{adsorption} & E^x \neq 0 \\ \text{surface reaction} & E^x_r = E^x{}_2 \\ \text{desorption} & E^x_3 \end{cases}$$

The energetic scheme may be more simple or more complex than above.

- Very simple case : $2HD \rightleftarrows H_2 + D_2$
  The only parameter is the adsorption heat $Q = E^x$ ;
  We expect the rate to increase when Q decreases ; in fact if Q is too small, the density of the adsorbed phase gets too low. These 2 opposite factors lead to a "volcano" shaped curve $v = f(Q)$.
  This is a rather general principle in catalysis : an optimal adsorption of the reactant is required.

- Complex cases. With heavy molecular reactants, the number of accessible adsorbed states becomes rapidly very large.
  Some of these states open catalytic reaction pathways, while other are dead-ends.
  Catalytic selectivity is high when one of these reaction pathways is strongly favored.
  A detailed kinetic analysis of these complex reaction networks, combined with isotopic tracing, lead to the structure of active intermediates.

For ex. This analysis has been pushed very far for the systems hydrogen-hydrocarbon, and the following adsorbed species have been proposed :

$C_nH_{2n+2}$

(-H) $C_nH_{2n+1}$

(-2H) $C_nH_{2n}$

(-3H) $C_nH_{2n-1}$

(-4H) $C_nH_{2n-2}$

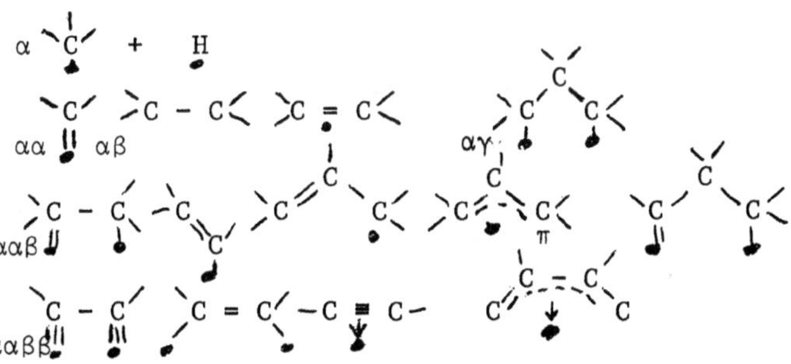


The mechanism of olefinic double bond hydrogenation and migration may be represented by scheme (Fig. 6)

$$C_3H_6 + H_2 \rightarrow C_3H_8$$
$$CH_2 = CH\text{-}CH_3 \rightleftarrows CH_3\text{-}CH = CH_2$$

On Ni, the rapid interconversion of $\alpha = \alpha\beta$ species leads to extensive migration ; this process is slow on $P_t$.

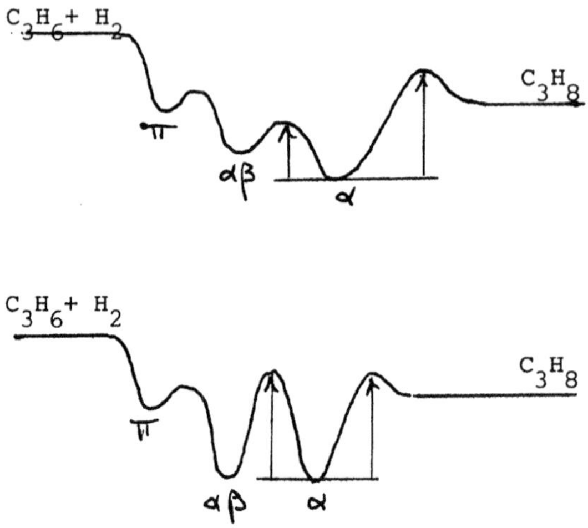


Fig.6

Ni

Pt

In fact, this scheme is very much simplified (according to F.G.Gault, 5 different mechanisms are possible for a simple cis-trans isomerization).

c) Conclusion :

A theory of catalysis would have a predictive value if all the accessible adsorption states of a given molecule on a given metal could be forecast. This is a formidable task.

As a first step, the prediction of the reactivity (Q, $E^x$) of a postulated species could be a very great help to test various hypothesis based on experimental results.

## II. REAL COMPLEXITY OF THE PROCESSES ON METALS.

Progress of kinetic and mechanistic analysis and new techniques of surface science (TPD, TPR, ESCA, AUGER, LEED....) revealed unexpected complexities of the elementary processes.

a) Adsorption.

Adsorbed states are frequently multiple ; adsorption heat decreases with surface coverage, and thermal desorption spectra may have several peaks.

Ex. (Fig.7) CO/W several discrete associative states
$H_2$/W several discrete dissociative states

Fig.7

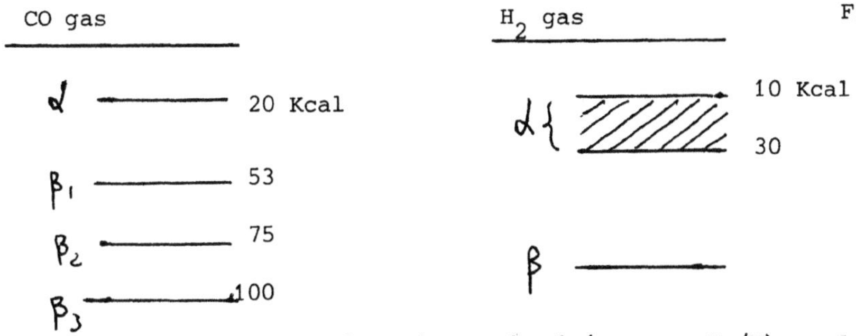


Some states are enlarged to a band (state, α $H_2$/W), a fact assigned to interactions between adsorbed particles ; but the existence of well separated states implies different adsorption sites (heterogeneity).

Models of usual faces, deduced from crystal structure, show that an ad-atome may occupy different positions and be bound to 1,2,3,4 atoms of metal, with energies $Q_1 < Q_2 < Q_3 < Q_4$ (Fig.8)

Fig.8

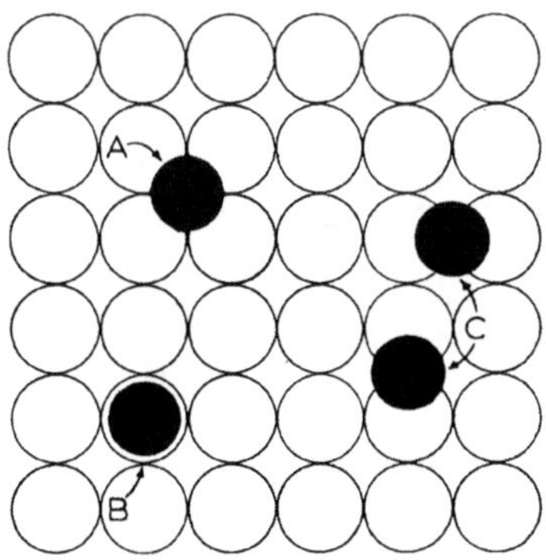


Moreover, the various faces of the same crystal present different symetries and therefore different sites : this point was experimentally demonstrated with large single crystals.

A direct determination of clean surface structures is now possible by Leed, with the following results :

- the theoretical structure of the crystal face is sometimes observed : $P_t$ (111) ; but in other cases, a reconstruction has taken place (fig.9) : $P_t$ (100)

Fig.9

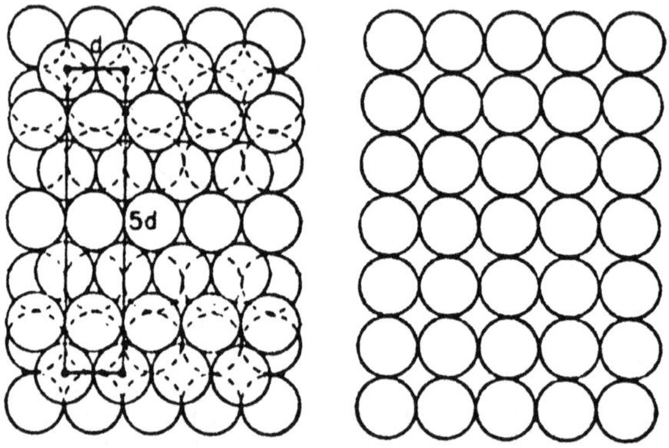


Reconstruction is the rule for high index faces, which are really formed of low index terraces separated by monoatomic steps :

$P_t$ (557) $P_t$ 9(111) x (100) (fig.10)

Fig.10

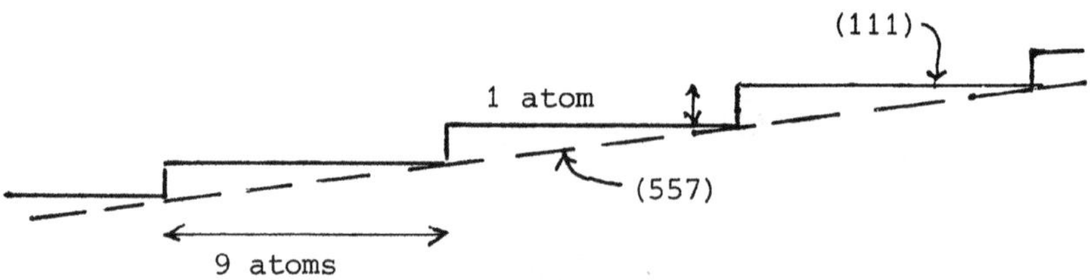


Direct observation by Leed during adsorption on a clean surface indicates that the ad-layer may be disordered or ordered depending on the nature of the adsorbate, the coverage of the surface, and the temperature :

CO/Pd (111) 3 ordered structures
$\theta$ : 0-1/3-1/2-2/3

If the heat of adsorption is high, a surface reconstruction takes place (Fig. 11) :

$O_2$/Ni (100) 2 structures
$\theta$ : 1/4 - 1/1

Fig.11

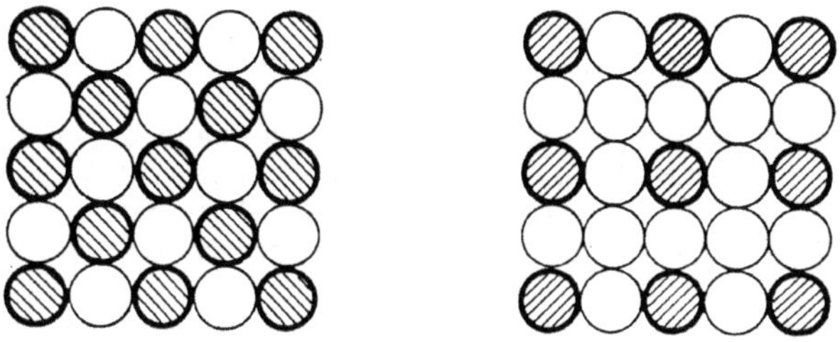

In extreme cases, new cristalline faces appear during adsorption :

$O_2$/W (112) → (110) (facetting)

The surface metal atoms are "mobilized" by adsorption.

b) Catalysis

- Specificity of crystal faces and steps.

For a given catalytic reaction rate and selectivity changes are influenced by surface structure in some reactions :

For ethylene hydrogenation on nickel, the (100) face is inactive : (110) > (111) ≫ (100)

In reaction of hydrocarbons with hydrogen on platinum, H-H and C-H bonds are broken on steps, C-C bonds on kinks, hence the high activity of vicinal faces when compared to low index planes.

- Effect of cristallite size.

For highly divided metals, catalytic properties may change with dispersion D (percentage of surface atoms), and two classes of reactions have been distinguished (Boudart).

| "Facile" or structure insensitive reactions (rate independent of D) | "Demanding" or structure sensitive reactions (rate change with D) |
| --- | --- |
| Hydrogenation (and reverse) | Hydrogenolysis (C–C) |
| Ethylene on Pt, Ni | Ethane on Rh |
| Benzene on Pt, Pd, Ni | Cyclopentane on Pt |

In fact, selectivity changes are easier to observe. Thus, in hydrogenolysis of methylcyclopentane on Pt at 260°C, the selectivity to n-hexane increases with dispersion.

These effects appear when the percentage of edge atoms become important, that is with crystal size of less than 50 Å.

But in this range, the usual crystal structure may not be stable, as indicated by calculations on clusters with less than 100 metal atoms.

Size effects are closely related to step-effects in mono cristal faces.

- Activity patterns.

Relationships between chemical nature of the metal and catalytic properties are frequently researched by plotting catalysis rates vs atomic number (or position on the periodic table). Transition metals are, as a rule, the most active catalysts (hence the importance of d-orbitals), and among them, the triads of the VIII$^{th}$ group. A dramatic fall of activity follows the d-band filling (at I$^{B}$ group : Cu, Ag, Au) for all hydrogenation and oxidation catalysis.

But unexplained differences are observed within the VIII$^{th}$ group metals for similar reactions.
Rh is the most active for ethylene hydrogenation,
Pt for benzene hydrogenation.

- We only mention here the interesting problems of alloys (or bimetallic) catalyst and supported metals (dispersed on carriers).

## III. PARTICULARITIES OF NON-METAL CATALYSTS.

The preceeding discussion was based on metals the most advanced field of experimental and theretical catalytic studies.

Generalization of principles and techniques to non-metal catalysts meets with a number of difficulties, frequently due to the limits of physical and chemical stability of multicomponent solid phases : the surface structure and composition and therefore the nature of active sites, may here change considerably with temperature and gas phase composition during adsorption or catalysis.

We shall discuss the case of metal oxides.

a) Oxides of multivalent metals.

As a rule, these metals form a series of oxides corresponding to different oxidation states :

$V_2O_3$ ($V^{III}$, $V_2O_4$ ($V^{IV}$), $V_2O_5$ ($V^V$)

and, for each one of these, a domain of non-stoechiometry (excess or defect oxygen).

The thermodynamic equilibria between the gas phase and the solid ($O_2$ chemical potential) regulates the structure of the catalyst, which is a function of temperature and redox potential (oxidizing or reducing atmosphere).

However, for kinetic reasons, this gas-solid equilibrium is not necessarily established. At low temperature, due to the slow diffusion rates in the solid, these equilibria are frozen. At $T > 0{,}3\ T_f$ ($T_f$, melting point K) superficial layers are equilibrated, and at $T > 0{,}5\ T_f$ bulk equilibrium is possible.

To summarize the real surface state is here hardly predictable, and the techniques of surface science (UHV) are difficult to apply.

Active centers include surface anions and cations, and lattice oxygen is frequently a potential reagent in catalytic reactions (redox processes).

b) Stable metallic oxides.

Oxides with only one stable oxidation state are extremely stable, and active centers are incompletely coordinated surface cations or anions acting as acidic (Lewis) or basic sites :

$$Al_2O_3 \quad : \quad Al \leftarrow \quad O \rightarrow$$

However these surfaces may react with water, and display a protonic acidity (Bronstedt) due to certain hydroxyl groups :

$$-OH \quad -O^{\ominus} + H^{\oplus}$$

The hydration modifies the density of both acidic and basic centers.

c) Bifunctional oxides.

It is easily seen that class a) oxides, with redox sites, also present the acid-base sites of class b) and may act as bifunctional catalysts.

Similar systems can be synthetized by deposition of a divided metal on an acidic support ($Pt/Al_2O_3$).

Both functions may be used successively in complex multistep catalytic reactions.

In other cases, the bifunctional behavior is undesirable and produces parasitic reactions.

In summary, solid oxides catalyst present, by nature, a large variety of active centers. Moreover, the type and density of these centers may change with the catalytic conditions (state of oxido-reduction or of hydration), and a reference state is not easily defined.

For there reason, this field of catalysis is so far the least understood except in particular cases (zeolites).

## IV. THE THEORIES OF CATALYSIS.

The various attempts to correlate all these facts in a general theory of catalysis have met, so far, with limited success, and have little predictive value. In the recent past, two types of approaches have been popular ; emphasizing respectively the collective or local properties of the solid catalyst :

- The "electronic theory of catalysis" ( 1945) considers the collective properties of the solid expressed by the band diagram, and the electronic transfer molecule.solid, determined by the Fermi level. This theory led to measure changes of electronic conductivity and work function during adsorption and catalysis on metals and semiconductors, to doping by particular impurities, and to photocatalysis.

- The theory of active centers considers only the binding site on the surface, and its nearest neighbours, exactly like a metallic complex in inorganic chemistry, to which ligand field theory is applied.

This theory emphasizes the coordination number, the symetry, the oxidation state, of the active center and their change by adsorption.

The relevant orbitals are simply predictable from the bonding and non-bonding orbitals existing in the crystal (surface valences).

Obviously, catalytic action is not entirely localized, and not entirely collective, and the models of the solid are oversimplified in both cases.

In recent years, an increasing number of attempts have been made to apply theoretical chemistry in our field : M-O calculations on metallic clusters, crystal field interpretation of EPR spectra of paramagnetic centers in oxides,...

We hope that this meeting will contribute to popularize this type of research among theoretical chemists to help us understand the many facets of surface catalytic action, keeping in mind that a general theory of catalysis is, at the present times, an overambitious goal.

## FURTHER READING

- Langmuir, I., Collected works (Suits, C.G., Rideal, E., Bridgman,P.W., & Way, H.E., editors). Pergamon Press N.Y. 1962.
Vol.1,3,8,9 (of historical interest as the first correct description of fundamental processes in surface catalysis).
- Clark, A., The theory of adsorption and catalysis. Academic Press N.Y. 1970 (a review of actual theories, mostly for adsorption).
- Ehrlich, G. Modern methods in surface kinetics (Flash desorption, Field emission Microscopy, ultra-high vacuum techniques), in Adv. in Catalysis. Vol.13 Academic Press N.Y.1963.
- May, J.W., Discovery of surface phases by Low Energy Electron Diffraction (Leed) ; in Adv. in Catalysis, Vol. 21, Academic Press, N.Y. 1970.
- Engel,T. & Ertl, G., Elementary steps in the catalytic oxidation of carbon monoxide on platinum metals, in Adv. in Catalysis, Vol. 28 Academic Press N.Y. 1979.
- Baetzold, R.C., Application of molecular orbital theory to catalysis in Adv. in Catalysis Vol 25 Academic Press N.Y. 1976.
- Ecole d'Eté du CNRS. Théorie des métaux et Catalyse. Lyon-Villeurbanne, Sept. 1975 (2 fasc.)
- Ecole de Printemps du CNRS, Aspects théoriques et pratiques de la Catalyse d'oxydation, Lyon, Mai 1978 (1 fasc.)
published as :
"Chemical and Physical Aspects of Catalytic oxidation.
Editions du CNRS. Paris 1980.

# ANALYSIS OF CO-METAL CLUSTER INTERACTION ENERGIES BY THE HARTREE-FOCK-SLATER METHOD

E.J. Baerends and D. Post
Scheikundig Laboratorium der Vrije Universiteit
De Boelelaan 1083
1081 HV Amsterdam, The Netherlands

ABSTRACT

The Hartree-Fock-Slater method is briefly described. Some small molecules serve to illustrate its performance with respect to several molecular properties. As examples of applications to the problems of coordination and adsorption we discuss the bonding of bridging COs in $Fe_2CO_9$, and the adsorption of CO at different sites of a Cu surface.

## 1. INTRODUCTION

The improvements in high vacuum technology and the development of a whole plethora of new experimental techniques for studying surfaces - UPS, XPS, LEED, ELS, SIMS etc. - has raised hopes of unravelling the functioning of heterogeneous catalysts. At the same time there is a growing interest in inorganic cluster complexes, both because of the challenge they provide to experimentalists and theoreticians, and because of the possibility that insight in chemisorption may be obtained from a study of the metal-ligand bond in the larger clusters, where the nucleus of metal atoms may be looked upon as a small metal particle [1]. It is yet an open question to which extent cluster complexes do indeed provide us with realistic models of the type of interactions that determine the adsorption processes. This question is in fact related to the long standing controversy on the localized versus delocalized nature of the surface chemical bond. That for some properties at least a quite striking analogy may be observed between cluster complexes of increasing size and surfaces has been nicely demonstrated in the UPS and XPS studies of Plummer et al. [2] on carbonyl complexes and chemisorbed CO.
Although of much interest, this particular area of research poses considerable difficulties for the theoretician. The sheer size of the systems is a serious obstacle for application of the standard ab initio Hartree-Fock programs of quantum chemistry, in particular if one requires the basis sets to be of double-zeta quality or better. It is of course possible to use some semi-empirical method instead. The work



*R. Daudel, A. Pullman, L. Salem, and A. Viellard (eds.), Quantum Theory of Chemical Reactions, Volume III, 15–33.*

by R. Hoffmann and coworkers [3] shows what tremendous insight may be obtained from such calculations. It is nevertheless desirable in many cases to have methods which are more reliable quantitatively. This holds true in particular for areas like surface chemistry where there is neither much previous experience to help develop the important "chemical intuition", nor experimental geometries and other data to ensure the right outcome of the theoretical "predictions".
There is thus an obvious need for new quantum chemical methods in the field of inorganic chemistry and catalysis. It is the purpose of the present contribution to indicate the potential usefulness of the so-called Hartree-Fock-Slater method [4,5,6]. It is clear that compared to ab initio Hartree-Fock calculations we are looking first of all for improvement in the speed of computation. This can only be achieved by using either a different one-electron model than HF, or different numerical techniques, or, of course, both. This should not, however, impair the reliability of the results.
As far as the numerical techniques are concerned, we require reasonable numerical accuracy (0.1 eV, say, in energies). As for the one-electron model to be chosen, it is reasonable to require similar agreement with experiment as can be obtained with Hartree-Fock. Of course the Hartree-Fock one electron model is only optimal in one sense (i.e. with respect to total energies of one-determinantal wavefunctions), and has well-known deficiences. It can therefore not be used to judge one-electron methods in general. Rather, its performance is often so poor that it would be undesirable to not even obtain the kind of agreement with experiment that Hartree-Fock offers. We wish to emphasize here that the ultimate criterion for any one-electron method is experiment, or the exact wavefunction, not Hartree-Fock as the most widely used one-electron model. Finally we may mention amongst the desiderata for any quantum chemical method for (very) large systems, the possibility of analyzing the enormous amount of numerical data (total and one-electron energies, basis function coefficients and populations) in meaningful concepts. As an example of the type of analysis we have in mind here we may mention the work by Morokuma et al. within the Hartree-Fock method [7].

The Hartree-Fock-Slater method to be discussed is based on the considerations mentioned above. The paper is organized in the following way. First we will briefly describe the essential features of the Hartree-Fock-Slater (HFS) scheme (section 2), which will be followed by some test results on simple molecules (section 3). As specific applications in the field of cluster complexes and chemisorption we will discuss the bonding of CO to metal atoms for the bridge CO in $Fe_2CO_9$ and for CO at different sites of a Cu crystal surface (section 4).

## 2. THE HARTREE-FOCK-SLATER METHOD FOR MOLECULES

We distinguish between the one-electron model to be used, and the computational techniques employed in order to obtain the one-electron wavefunctions corresponding to this model.
The Hartree-Fock-Slater model differs from Hartree-Fock in the use of Slater's local $\rho^{1/3}$ exchange-correlation potential:

$$V_{X\alpha}(1) = -3\alpha\left[(3/8\pi)\ \rho(1)\right]^{1/3}$$

The constant $\alpha$ in this potential has always been given the value 0.70 [4b]. The use of the Kohn-Sham-Gaspar value of 2/3 would give negligible differences.
The one-electron equations:

$$\begin{aligned}\hat{h}_{HFS}\ \psi_i(1) &= \{-\tfrac{1}{2}\nabla^2(1) + V_N(1) + V_C(1) + V_{X\alpha}(1)\}\psi_i(1) \\ &= \varepsilon_i\ \psi_i(1) \qquad (1)\end{aligned}$$

may be derived variationally from the so-called statistical total energy:

$$\begin{aligned}E_{X\alpha} &= \sum_i n_i <\psi_i \left| -\tfrac{1}{2}\nabla^2 + V_N \right| \psi_i> \\ &+ \tfrac{1}{2}\int V_C(1)\ \rho(1)\ d\tau_1 + \tfrac{3}{4}\int V_{X\alpha}(1)\ \rho(1)\ d\tau_1 \qquad (2)\end{aligned}$$

The potential usefulness of the Hartree-Fock-Slater model is connected with the purely local character of the potentials appearing in the hamiltonian $\hat{h}_{HFS}$. This makes it possible to use direct numerical integration techniques for evaluating integrals where $\hat{h}_{HFS}$ appears in the integrand. This fact has been exploited in a computational scheme, the main features of which may be summarized in the following points:

a) The orbitals $\psi_i$ are expanded in a basis set of exponential functions (STOs):

$$\psi_i = \sum_\mu \chi_\mu\ C_{\mu i} \qquad (3)$$

The primitive functions (or symmetry combinations thereof) are orthonormalized by a symmetrical or canonical orthonormalization, in which the required overlap integrals are evaluated analytically. It is well known that minimal basis sets lead to quite distorted results for many molecular properties. We therefore always use basis sets of double zeta quality, and often better. For an investigation of basis set effects in transition metal complexes we refer to [8]. There is a possibility to keep the cores frozen, in which case the valence basis set is augmented with a single zeta set of core functions which are used to orthogonalize the valence set on the cores [4a]. Full double (or triple etc.) zeta flexibility for the valence orbitals is thus preserved.

b) Expressing the $\psi_i$ in the orthonormalized basis functions, the matrix elements in the resulting matrix equation:

$$\underline{\underline{H}}\, \underline{c}_i = \varepsilon_i\, \underline{c}_i \tag{4}$$

are evaluated by a numerical integration procedure:

$$H_{\mu\nu} = \langle \chi_\mu | \hat{h}_{HFS} | \chi_\nu \rangle$$

$$\approx \sum_k \chi_\mu^*(\bar{r}_k)\, \hat{h}_{HFS}\, \chi_\nu(\bar{r}_k)\, w(\bar{r}_k) \tag{5}$$

The sample points $\bar{r}_k$ and weight factors $w(\bar{r}_k)$ are determined according to the Diophantine method introduced by Ellis and Painter [9]. The integrals are not determined very accurately by this procedure, as we use only a moderate number of sample points (500 - 1000 per atom). In many cases however an accuracy of e.g. ∿0.1 eV in the hamiltonian matrix elements is quite sufficient. In fact the "error bars" that one should apply to calculated properties such as bond energies, ionisation potentials etc. are often considerably larger than this due to basis set deficiencies. In general it is important to balance the various approximations one is using, a point to which we will return below.

c) A numerical integration is possible since we have only local potentials - $V_N$, $V_C$ and $V_{X\alpha}$ - in the hamiltonian. The calculation of $V_N$ and $V_{X\alpha}$ in each sample point is very straightforward. Exact calculation of $V_C$:

$$V_C(\bar{r}_k) = \int \frac{\rho(r)}{|\bar{r}-\bar{r}_k|}\, d\bar{r} = \sum_{\mu,\nu} P_{\mu\nu} \int \frac{\chi_\mu(\bar{r})\, \chi_\nu(\bar{r})}{|\bar{r}-\bar{r}_k|}\, d\bar{r} \tag{6}$$

would require the evaluation of many three-center nuclear attraction integrals. We simplify the calculation of $V_C(\bar{r}_k)$ by expanding the electron density in a set of functions (STO's) centred on the nuclei of the system:

$$\rho(1) \approx \sum_i a_i\, f_i(1) \tag{7}$$

The coefficients $a_i$ are determined by a least squares fitting procedure [4a]. At present we use as the fitting functions all one-center products of basis functions, except when this would lead to linear dependency problems. With such a set of fitting functions the density is described very accurately. This is an important difference with the MSXα method, where the muffin-tin approximation is applied to the potentials $V_C$ and $V_{X\alpha}$. The Hartree-Fock-Slater (or Xα) model however yields quite good agreement with experiment for many molecular properties only if the full non-muffin-tin potentials are used [5,10].

d) As a general principle we try to maintain a certain balance between the approximations introduced at several stages. For example, given the fact that orbital energies (and ionization potentials) can not be trusted to more than a few decimal places ($\sim$0.01 a.u.) in a double-zeta basis set calculation, it is a waist of time to compute them to the full machine accuracy of, say, 16 digits. We therefore cut down the computation time of the hamiltonian matrix by using a simple $n^2$-dependent numerical integration (n is the number of basis functions). On the other hand, much of the reliability and usefulness of the Hartree-Fock-Slater model is lost if the electronic potentials are not treated accurately, so we spend a comparatively large amount of time on computing them accurately.

The calculation of total energies is a special problem in the scheme described above. As we are in general interested in small energy differences, such as the chemisorption energies of small molecules on clusters, the total energy has to be calculated very accurately. It is rather time-consuming to calculate a term like $\int V_C(1)\rho(1)d\bar{r}_1$ to the required high accuracy using the same integration scheme we use for the hamiltonian matrix elements.
An elegant solution to this problem has been proposed by Ziegler and Rauk [11], who extended Slater's transition state method to the calculation of bond energies. We will briefly review this method as we will also use it for an analysis of chemisorption energies.
When using an expansion method we may write the spin polarized statistical total energy in terms of integrals over the basis functions:

$$E_{X\alpha} = \sum_{\sigma}\sum_{\mu,\nu} P^{\sigma}_{\mu\nu} \langle\chi_\mu|-\tfrac{1}{2}\nabla^2 + V_N|\chi_\nu\rangle + \sum_{\sigma,\tau}\sum_{\mu,\nu}\sum_{\kappa,\lambda} P^{\sigma}_{\mu\nu} P^{\tau}_{\kappa\lambda} \langle\chi_\mu(1)\chi_\kappa(2)|\frac{1}{r_{12}}|\chi_\nu(1)\chi_\lambda(2)\rangle + \sum_{\sigma} K\int\Big[\sum_{\mu,\nu} P^{\sigma}_{\mu\nu}\chi^{*}_{\mu}(1)\,\chi_\nu(1)\Big]^{4/3} d\tau_1, \qquad (8)$$

with $K = -\frac{9}{4}\alpha\,(\frac{3}{4\pi})^{1/3}$

Here $\sigma$ (and $\tau$) run over the spin up and down possibilities, and $P_{\mu\nu}$ are the elements of the charge and bond order matrix. Straightforward differentiation shows:

$$\frac{dE_{X\alpha}}{dP^{\sigma}_{\mu\nu}} = \langle\chi_\mu|-\tfrac{1}{2}\nabla^2 + V_N + V_C + V^{\sigma}_{X\alpha}|\chi_\nu\rangle = F^{\sigma}_{\mu\nu} \qquad (9)$$

If we make a special choice of basis functions, namely the SCF orbitals of the system, the P-matrix will be diagonal and the diagonal elements of the F-matrix the one-electron energies. Equation (9) reduces to

$$\frac{dE_{X\alpha}}{dn_i} = \varepsilon_i \tag{10}$$

which was the basis of Slater's transition state method for ionization and excitation energies. We consider the case of a system which is build up from the fragments A and B. It is then convenient to use as basis-functions the (occupied and virtual) SCF orbitals of the separate fragments, $u_i^A$ and $u_j^B$. In this basis set the superposition of the densities of A and B obtained by bringing the unmodified fragments together, is described by a diagonal P matrix. This situation, denoted by AB, will be taken as a starting point:

$$\rho_{AB} = \rho_A + \rho_B = \sum_\sigma \sum_i P_{ii}^\sigma u_i^* u_i \tag{11}$$

$$E_{AB} = E_A + E_B + \Delta E_{elstat} + \Delta E_{exch} + \Delta E_{exch} \tag{12}$$

Here $E_A$ and $E_B$ are the energies of A and B separately, $E_{AB}$ is obtained by simply substituting into eq. (8) the $P_{ii}$ and $u_i$ of eq. (11), $E_{elstat}$ is the electrostatic interaction between A and B when brought from infinity to their final positions:

$$\Delta E_{elstat} = \sum_\alpha^A \sum_\beta^B \frac{Z_\alpha Z_\beta}{R_{\alpha\beta}} + \int \frac{\rho_A(\bar{r}_1)\rho_B(\bar{r}_2)}{r_{12}} d\bar{r}_1 d\bar{r}_2 - \sum_\alpha^A \int \frac{Z_\alpha \rho_B(\bar{r}_1)}{|\bar{r}_1 - \bar{R}_\alpha|} d\bar{r}_1 - \sum_\beta^B \int \frac{Z_\beta \rho_A(\bar{r}_1)}{|\bar{r}_1 - \bar{R}_\beta|} d\bar{r}_1, \tag{13}$$

and $\Delta E_{exch}$ is the difference in statistical exchange energies between a system with density $\rho_A + \rho_B$ and the separate systems A and B:

$$\Delta E_{exch}^\alpha = K\int (\rho_A^\alpha + \rho_B^\alpha)^{4/3} d\tau - K\int (\rho_A^\alpha)^{4/3} - K\int (\rho^\alpha)^{4/3} d\tau. \tag{14}$$

We note that $E_{AB}$ is not a physically meaningful quantity, but $E_{X\alpha}$ of eq. (8) is used here purely formally as a function of the $P_{\mu\nu}$, given the basis set. After a complete SCF calculation on the total AB we will obtain the final density, $\rho_{\overline{AB}}$:

$$\rho_{\overline{AB}} = \sum_\sigma \sum_{i,j} (P_{ii}^\sigma \delta_{ij} + \Delta P_{ij}) u_i^*(1) u_j(1) \tag{15}$$

Using a Taylor expansion of both the "initial" energy $E_{AB}$ and the final energy $E_{\overline{AB}}$ around a transition state defined by

$$\rho_{TS} = \tfrac{1}{2} (\rho_{\overline{AB}} + \rho_{AB}) \tag{16}$$

it can be shown [11] that to fourth order in $\Delta P_{ij}$

$$E_{\overline{AB}} - E_{AB} = \sum_{i,j} \left[ \frac{1}{6} \left(\frac{dE}{dP_{ij}}\right)_{AB} + \frac{2}{3} \left(\frac{dE}{dP_{ij}}\right)_{TS} + \frac{1}{6} \left(\frac{dE}{dP_{ij}}\right)_{\overline{AB}} \right] \Delta P_{ij}$$

$$= \sum_{i,j} \left[ \frac{1}{6} F_{ij}^{AB} + \frac{2}{3} F_{ij}^{TS} + \frac{1}{6} F_{ij}^{\overline{AB}} \right] \Delta P_{ij}$$

$$= \sum_{i,j} F_{ij}^{eff} \Delta P_{ij} \tag{17}$$

Combining equations (12) and (17) we obtain the bonding energy between the fragments A and B as:

$$\Delta E = E_{\overline{AB}} - E_A - E_B = \Delta E_{elstat} + \Delta E_{exch} + \sum_{i,j} F_{ij}^{eff} \Delta P_{ij} \tag{18}$$

The usefulness of this equation stems from the fact that it allows the determination of the quantity we are interested in - the bonding energy - directly, without subtraction of two total energies. The first two terms, $\Delta E_{elstat}$ and $\Delta E_{exch}$ are calculated using the expansion of the density in one-center fitting functions (eq. 7), thus obviating the need to evaluate three- and four-center integrals. As we are, in general, not dealing with interpenetrating fragments, the quality of the fit is not crucial. The Fock matrix in the last term is calculated in exactly the same way as during SCF iterations.

Apart from having the advantage of high numerical stability, the transition state expression for the bonding energy is also very convenient for purposes of analysis. We will differentiate here between the occupied-occupied orbital interactions (exchange repulsion) and the occupied-virtual orbital interactions (charge-transfer, polarisation). When constructing a properly antisymmetrized wavefunction

$$\Psi_{\widetilde{AB}} = | u_1^A \ldots u_N^A u_1^B \ldots u_M^B | \tag{19}$$

from all occupied orbitals on fragments A and B, the corresponding density $\rho_{\widetilde{AB}}$ (obtained e.g. by mutually orthogonalizing the sets $\{u_i^A\}$ and $\{u_j^B\}$) may be used in the transition state method to obtain the steric interaction energy (electrostatic interaction plus exchange repulsion)

$$\Delta E^0 = E_{\widetilde{AB}} - E_A - E_B \tag{20}$$

(N.B. $E_{\widetilde{AB}}$ should not be confused with the physically meaningless $E_{AB}$!) The energy change in going from $\Psi_{\widetilde{AB}}$ to the converged $\Psi_{\overline{AB}}$ will be given by an expression of the form

$$E_{\overline{AB}} - E_{\widetilde{AB}} = \sum_{i,j} F_{ij}^{eff} \Delta P_{ij} = \sum_{\Gamma} \Delta E^{\Gamma} \quad (21)$$

where the $\Delta P_{ij}$ reflect the change from $\rho_{\widetilde{AB}}$ to $\rho_{\overline{AB}}$. As the Fock-matrix is symmetryblocked we will get energy contributions corresponding to each irreducible representation $\Gamma$ of the symmetry group used. Very often the symmetry is such that $\sigma$- and $\pi$-bonds fall in different irreducible representations. Expression (21) then allows a very natural distinction between the energy contribution (or strength) of such bonds (cf applications of this in refs. [12] and [13]). This is very valuable, as the inorganic chemistry, and more recently the surface science literature, are replete with speculations on the relative importance of e.g. $\sigma$-donation and $\pi$-backdonation in coordination and adsorption of CO. The opinions range from very minor contribution of the $\pi$-bond to virtually exclusive $\pi$-bonding, so a quantitative assessment is very interesting indeed.

## 3. APPLICATIONS TO SIMPLE DIATOMIC MOLECULES

Before we proceed to a discussion of applications of the HFS method on larger systems, we wish to build up some feeling for the type of agreement with experiment that can be obtained with the method.
If one wants to know what the capabilities of the HFS model are, it is necessary to minimize computational approximations. We have done calculations on a series of diatomic molecules in which the numbers of sampling points in the numerical integration, and the number of fit functions in the density fit, have been chosen so large that the results of the calculations were completely determined by the basis set chosen, as in ab initio HF calculations. In fig. 1 we show the deviation of the dissociation energies obtained in a large basis set (triple zeta s and p plus polarization functions for the first row elements [14], triple zeta 3d and double zeta 4s and 4p for Cu) from the experimental values. The HFS results are compared to the best HF results available (cf ref. [5]). Recently, Dunlap et al. [16] have obtained HFS results in essential agreement with ours, using an independently developed computational scheme. Their method is based on Gaussian basis functions and uses fits of both the charge density and the exchange potential by sets of auxiliary one-center functions. The close agreement between the results obtained with the different methods testifies to the numerical accuracy and basis set convergence of the calculations. A most remarkable conclusion we may draw from these HFS calculations is that there is a significantly better agreement with experimental dissociation energies than given by HF calculations.
The reason for this is not completely clear. The one-electron densities obtained in HF and HFS calculations are extremely close [17], and in general close to the exact one-electron density. There have been early

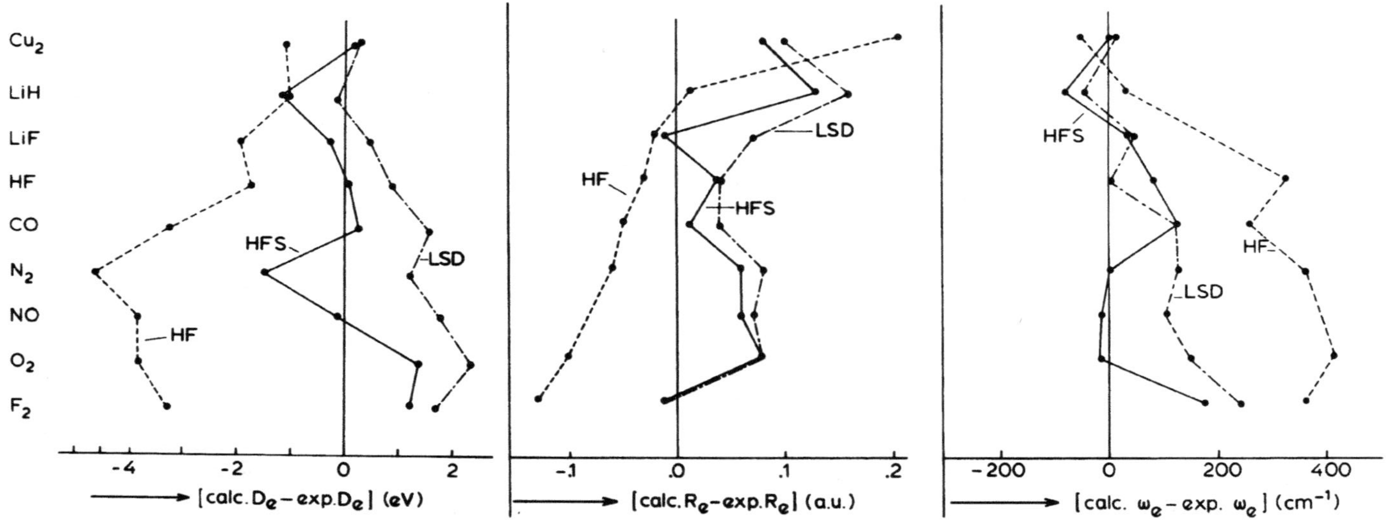


Fig. 1a 1b 1c

The deviation of calculated spectroscopic constants from experimental values, for both Hartree-Fock-Slater calculations (———), ab initio Hartree-Fock calculations (.....), and local (spin) density calculations (-·-·-).
For references see [5] and [26].
a) Dissociation energies
b) Equilibrium bond lengts
c) Harmonic stretch frequencies

attempts to obtain correlation corrections to the HF total energy from a knowledge of the HF one-electron density [18,19], recently revived [20] under the stimulus of the Hohenberg-Kohn-Sham theorem [21,22]. The use of the energy expression $E_{X\alpha}$ may be looked upon as a similar attempt, although it should be added that the reason for its success remains somewhat obscure. This is evident from the fact that the use of the local-(spin-)density functional expression for the exchange-correlation part of the energy [23], constructed to describe both exchange and correlation contributions, and therefore expected to be an improvement over the simple $X\alpha$ expression, does not lead to any improvement in the dissociation energies (see fig.1a). We may nevertheless conclude from our results that as far as calculations at the SCF level are concerned the HFS model is a good alternative to Hartree-Fock. This is further substantiated in figs. 1b and 1c, where the equilibrium bond lengths and vibration frequencies of the same series of diatomic molecules are compared with experiment and with HF and local-density functional results.

## 4. APPLICATIONS TO CLUSTERS AND SURFACES

One of the most interesting aspects of adsorption of a molecule like CO, compared to coordination, is the possibility of interactions of the adsorbate molecule with more than one metal atom (2 for a bridge site, 3 or more in a hollow site). In fact, much of the discussion of chemisorption phenomena revolves around the problem of establishing the preferred adsorption site. In the large group of transition metal carbonyl clusters, ranging from simple binuclear and trinuclear ones like $Fe_2CO_9$ and $Fe_3CO_{12}$ to such formidable structures as $Pt_{30}CO_{60}^{2-}$, we may find models for the on top, bridged and hollow surfaces sites in terminal, $\mu_2$- and $\mu_3$- bonded COs. We will first consider briefly the electronic structure of the bridging COs in $Fe_2CO_9$, and then proceed to a discussion of CO adsorbed at a top and a hollow site on a transition metal surface.

### $Fe_2CO_9$

In order to understand the electronic structure of the well-known CO-bridged binuclear complex $Fe_2CO_9$ it is important to note the nearly octahedral surrounding of each Fe nucleus by CO ligands (see fig. 2). This will lead to a splitting of the metal d states into a $t_{2g}$ set of d orbitals pointing in between the ligands, and an $e_g$ set having maximum density in the direction of the ligands. It has been assumed for a long time that a direct Fe-Fe bond exists in this system, because of its diamagnetic behaviour and in agreement with the 18-electron rule. In principle such a direct Fe-Fe bond might arise from interaction of a $t_{2g}$ orbital on one Fe, with highest density along the binuclear axis (if we take the Fe-Fe axis as z-axis $d_{z^2}$ is such a $t_{2g}$ orbital, see fig. 2a), with its partner on the other Fe nucleus. Bond formation implies a filled bonding and an unoccupied antibonding combination.

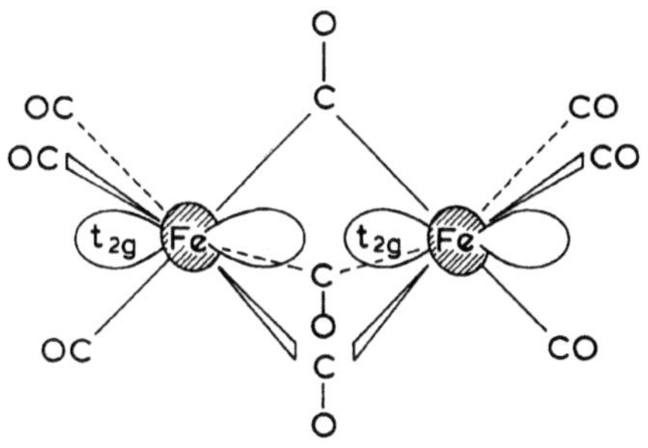


Fig. 2a

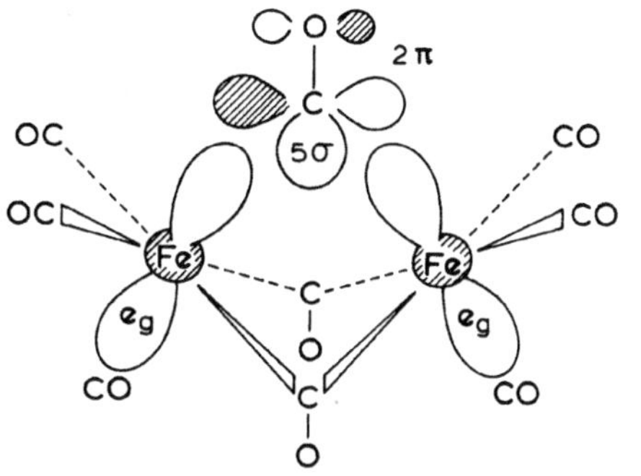


2b

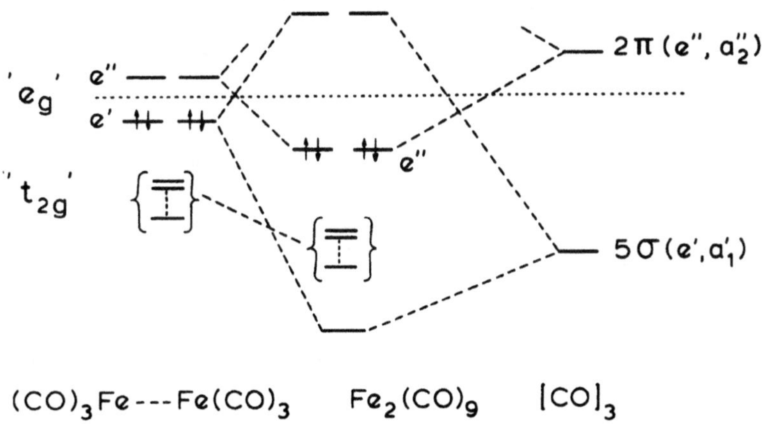


2c

a) The '$t_{2g}$' orbitals (in the local octahedral environment) in $Fe_2(CO)_9$ with maximum density along the Fe-Fe axis.
b) The $e_g$ orbitals in the local octahedra.
c) Schematic orbital interaction scheme for the bonding of the bridge COs to a $Fe_2(CO)_6$ fragment. The symmetry designations are according to $D_{3h}$. For details see [24].

As the $t_{2g}$ set of orbitals is relatively low-lying because of the ligand-field splitting induced by the π-acceptor CO ligands, this requires a large destabilization (larger than the ligand field splitting) of the antibonding combination. It turns out, by actual calculation, that the interaction between the $t_{2g}$ orbitals on the two Fe centers is not large enough. It is in fact fairly small. We give in fig. 2c a schematic representation of the important orbital interactions (see also ref. [24]). If we first bring two $Fe(CO)_3$ fragments together to form $Fe_2(CO)_6$, we find little splitting of the $t_{2g}$-like orbitals. In a $d^8$ atom like Fe the $t_{2g}$ set is completely occupied, and no net bonding will result. The $e_g$ - derived orbitals, the $d_{z^2}$ component of which is drawn in fig. 2b, interact more, and a bonding combination of these would be the highest occupied orbital in $Fe_2(CO)_6$. Note that we are now dealing with a "real" z-axis in the local octahedral environment of Fe, i.e. one which points to the ligands, not in between. The different behaviour of the $t_{2g}$ - derived "$d_{z^2}$" (fig. 2a) and the $e_g$-derived $d_{z^2}$ (fig. 2b) reflects the difference in spatial extent caused by interaction with the ligands [3,24]. When we switch on the interaction with the bridging COs, the antibonding combination of the $e_g$ orbitals (e" in the $D_{3h}$ symmetry of the complex) will be strongly stabilized by interaction with the $\pi^*$ orbitals of the bridging COs. The bonding combination e', is destabilized by interaction with the CO $5\sigma$ orbitals. The result is that the e' and e" levels cross, and the highest occupied orbital in $Fe_2CO_9$ is an e" orbital which is metal-metal antibonding. The absence of a direct metal-metal bond in this system is interesting, and we refer to [24] for a more detailed discussion, and comparison with other binuclear systems such as $Mn_2(CO)_{10}$ and $Co_2(CO)_8$ (see [27] for a discussion of the absence of direct metal-metal bonds in the carbonyl bridged metal-metal linkages in $Fe_3(CO)_{12}$ and $Fe_2Cp_2(CO)_4$). Here we draw attention to the way the bridging CO is bound to the $Fe_2$ "substrate". We have depicted this bonding as essentially taking place through a $(d_{z^2}(e_g) + \pi^* - d_{z^2}(e_g))$ three-center bonding orbital. It is clear that the geometrical arrangement of ligands around the Fe atoms is crucial for this bond: the ligand-field destabilized $e_g$ orbitals on Fe are close in energy to the bridge $\pi^*$, and have the right directionality to overlap strongly with them. This points to a very important difference with the chemisorption case. If the two Fe atoms involved in the bridge bond are surface atoms of a crystal, the energy and directionality of the "dangling bond" orbitals at the surface will depend upon the interaction with the atoms surrounding the surface in the first (top) and second layer. Not only the geometric positioning of these surrounding atoms, but also the fact that they are metal atoms instead of the usual ligands will be very important. The importance of the difference between $\pi$-acceptor (e.g. CO) and $\pi$-donor (e.g. Cl) ligands in a $ML_3$ fragment discussed recently by Summerville and Hoffmann [25] is very revealing in this respect. Moreover, we have not yet considered the possible role of the 4s and 4p atomic orbitals. In the discussion above we only took into account the ligand field splitting amongst the d-orbitals, and this is justified in view of the composition of the final molecular orbitals [24]. Nevertheless, in metals the sp band is known to be partly filled, and the relative contribution of conduction band and d band remains an interesting problem. We will discuss this in somewhat more detail for the case of CO adsorption on Cu.

$Cu_5$-CO

As a preliminary to an investigation of the electronic factors that are important in determining the chemisorption sites and the strength of the surface chemical bond we consider small clusters (five atoms in this case) which have been chosen so as to represent the immediate neighbourhood of the adsorbate (see fig. 3). The "top cluster" consists

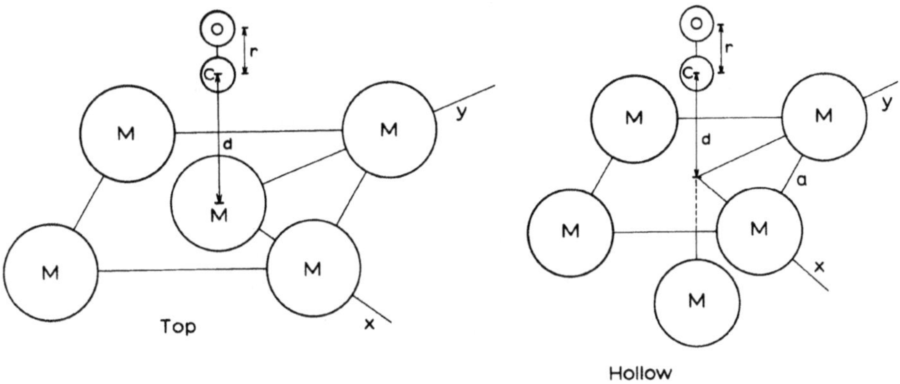


Fig. 3a 3b

Geometries of the five-atom clusters used to model chemisorption at a top (a) and hollow (b) site of the (100) surface of Cu crystal. The nearest neighbour distance is 4.8 a.u.

of the metal atom to which CO approaches and its 4 neighbours in the surface plane, the "hollow cluster" consists of the 4 atoms in the surface plane surrounding the hollow site and the metal atom in the next layer right under the hollow site. It is not to be expected that we will obtain a full understanding of the chemisorption phenomenon from such small clusters, let alone quantitative agreement with experiment, but they may be used to pin down the essential differences between e.g. a top and a hollow site, and to direct our attention to those size-dependent features of the electronic structure of clusters that are important for the chemisorption process. The choice of CO as adsorbate is not only motivated by the possible analogy (or contrast) with coordination in carbonyl clusters, but also by the possibility of probing both the σ-donation and π-backdonation types of interaction by this molecule. The symmetry of the clusters chosen ($C_{4v}$) reflects the local symmetry of the surface sites under consideration. In this symmetry the σ and π interactions occur in different irreducible representations ($A_1$ and E resp.), and can be split using expression (21). We will study the separate terms in the chemisorption energy as a function of surface-adsorbate distance, and of chemisorption site.

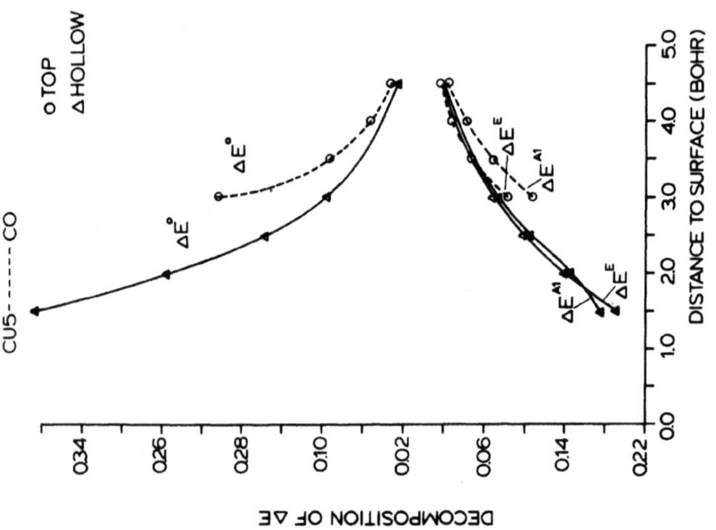


Fig. 5. Decomposition of the bonding energy of CO to a $Cu_5$ cluster. $\Delta E^o$ is the steric repulsion, $\Delta E^{A1}$ is the σ-bonding energy, and $\Delta E^E$ the π-(back)bonding energy.

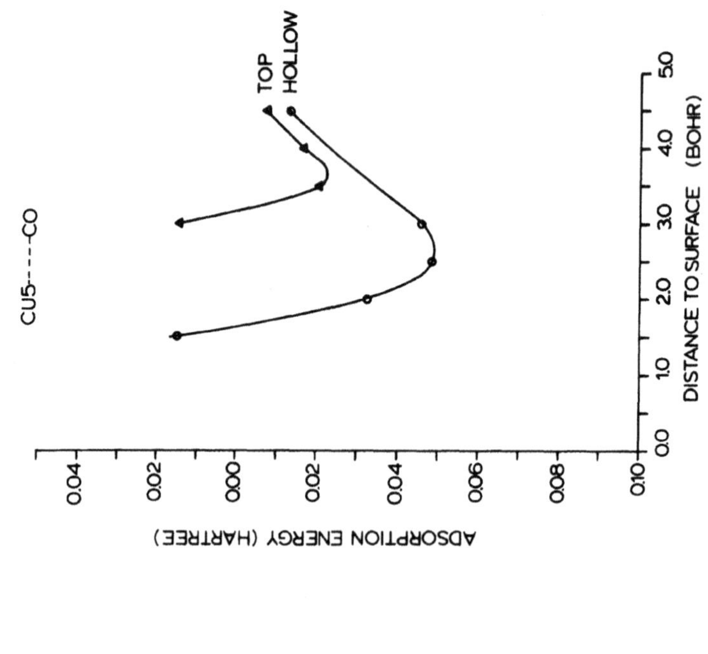


Fig. 4. The variation of the bonding energy of CO to a $Cu_5$ cluster with height above the surface (d in fig. 3).

In fig. 4 the bonding energy for a CO molecule to a $Cu_5$ cluster (both the top and hollow ones) is given as a function of the distance d. The C-O distance has been kept fixed at 2.14 a.u. At a very large distance from the surface the two curves coincide, but at shorter distances the "structure" of the surface plays a role. It is clear that in the top case the equilibrium distance will be at a much larger surface-adsorbate separation, with at the same time a considerable lower adsorption energy. The larger distance at the top site may be expected intuitively, but we will try to understand it, and the difference in chemisorption energy, from an analysis of the energy along the lines sketched in section II. In fig. 5 we show the different contributions to the bonding energy; $\Delta E^{O}$ is the steric interaction (electrostatic interaction plus exchange repulsion); $\Delta E^{A1}$ is the contribution from interactions in the A1 symmetry block, mostly the CO 5σ with virtual $a_1$ orbitals of the clusters (σ-donation); $\Delta E^{E}$ is due to interactions in the E symmetry, primarily involving the empty CO 2π (i.e. $\pi^*$) orbital and occupied metal states (π-backdonation). As a first point we note that the terms $\Delta E^{E}$ and $\Delta E^{A1}$ are roughly similar, which means that σ-donation and π-backdonation are about equally important. This is the case at both sites. At the top position there may be backbonding from occupied d orbitals on the central Cu atom, and from an e combination of 4s orbitals on the surrounding Cu atoms. On moving to the hollow site, the backbonding is mainly from an e combination of 4s orbitals, but this is stronger now due to favourable overlap of the $\pi^*$ orbital with this 4s combination. In fig. 6

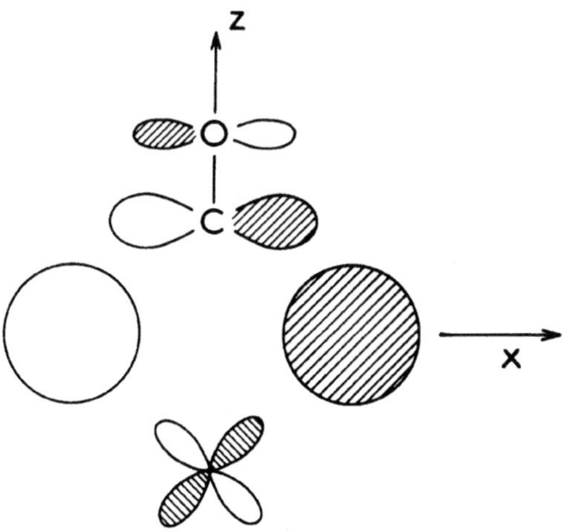


Fig. 6.
Schematic drawing of the overlap of the CO $\pi^*$ orbital with the highest occupied $Cu_5$ orbital, 7e. The 7e orbital is mostly localized on the surface atoms (predominantly 4s-4s), with small 3d admixture on the sub-surface Cu atom.

the favourable overlap between the highest occupied cluster orbital (7e) and $\pi^*$ orbital is illustrated. The net effect is that there is very little change in the electronic interaction terms in moving from a top to a hollow site (cf. the height of 3.10 a.u., for which both the top and hollow data are shown in fig. 5). As a corollary, the difference between the sites is largely determined by the steric interaction. From fig. 5 we see that the steric term becomes repulsive at much larger

distance from the surface in the top position than in the hollow position. This is due to the fact that the overlap of occupied CO orbitals with occupied orbitals of the central Cu atom becomes large, leading to strong exchange repulsion. It is to be noted that core and subvalence levels, which can be ignored for the bond formation, largely determine the exchange repulsion. They give the surface its hardspheres like appearance, not the electronic interaction terms. As the steric interaction starts its strongly repulsive upwards curvature closer to the surface in the hollow case, and as the electronic interactions become more strongly attractive on closer approach to the surface due to enhanced overlaps and interaction matrix elements, we find in the hollow case a shorter distance to the surface and a larger bonding energy.
It is interesting to compare CO in the hollow position with the bonding of bridge CO in $Fe_2CO_9$, and CO coordination in mononuclear complexes. In $Fe_2CO_9$ the backbonding to the bridge COs is still due to d orbitals, be it the $e_g$ instead of the $t_{2g}$ orbitals. In $Cu_5$-CO (hollow) the backbonding is mainly due to the electrons in an s state. This is an interesting difference with the case of coordination to one metal atom, as in carbonyl complexes. A copper carbonyl complex is not known, and would indeed be expected to be unstable because of the extra 4s electron outside the d shell. In the complex this extra electron would have to occupy an antibonding M4s-5σ(CO) orbital, or an antibonding Mdπ-$\pi^*$(CO) orbital. At a surface, however, there are different symmetry rules, and the possibility that 4s orbitals become involved in the backdonation is a very important factor in the stability of the chemisorption bond.

It is interesting to consider the effect of changes in the C-O bond length. In fig. 7 the dissociation energy curves are given for free CO and adsorbed CO (top and hollow). The free CO curve refers to the energy $\Delta E = E(CO) - E(C) - E(O)$, whereas the adsorbed CO curves refer to the energy $E(Cu_5CO) - E(Cu_5) - E(C) - E(O)$. The energy lowering from free CO to the adsorbed COs thus reflects the chemisorption energy. We note that in both the top and hollow cases the chemisorption energy becomes larger when we stretch the CO bond, this effect being more pronounced in the hollow case. The energy curves become more shallow, which will experimentally correspond to a decrease in C-O stretch vibration frequency, and in an elongation of the C-O bond. An explanation may be obtained from an analysis of the chemisorption energy in terms of steric interaction and σ- and π-bond formation as a function of the C-O internuclear distance. This analysis is represented graphically in fig. 8. It is clear that the σ-bond formation ($\Delta E^{A1}$) is not very sensitive to the C-O bond length in either position (top or hollow). The steric term is not completely independent of R(C-O), but the dominant factor in the change of adsorption energy when the C-O bond is stretched is evidently the π-backdonation $\Delta E^{E}$. The rationale for this is as follows. Stretching the C-O bond has little influence on the energy of the 5σ level, which has only little antibonding character, but it has a considerable stabilizing effect on the position of the strongly antibonding $\pi^*$ level of CO. The HOMO-LUMO interaction (CO $\pi^*$ - cluster 7e) is accordingly enhanced. As this interaction is more important at the hollow site than at the top site, the decrease in vibration frequency is calculated to

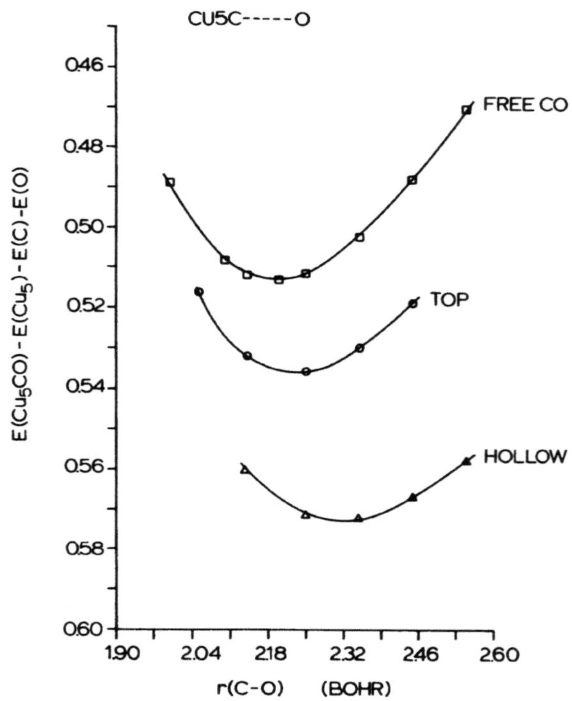


Fig. 7. Variation of the total energy with the C–O bond length. The energies of the C and O atoms, and the free $Cu_5$ clusters are substrated.

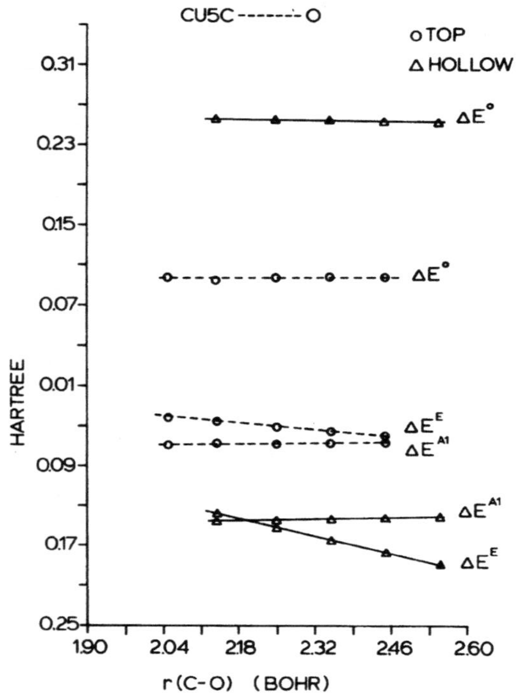


Fig. 8. Decomposition of the bonding energy of CO to a $Cu_5$ cluster as a function of the C–O bond length.

be largest at the hollow site.
Finally a comment on the comparison of our calculated chemisorption energies and geometries with experiment may be in order. We have spoken freely about "chemisorption energy" etc., but of course we really only calculated the interaction of small clusters with CO. It is our purpose to build up a detailed understanding of the interaction of prototype adsorbates like CO (and NO, $C_2H_4$, $C_2H_2$ etc.) with such clusters, in particular with respect to the differences they exhibit with the well-known bonding in organometallic systems. A full understanding of the chemisorption phenomenon will, however, require a knowledge of the effect of enlarging the clusters to the limit of a semi-infinite crystal on the important interaction terms. As far as our present clusters are concerned, we have found excellent quantitative agreement with the experimental data (heat of adsorption, surface-CO distance) for the top cluster. This may be fortuitous because this implies that the hollow cluster gives a too large bonding energy, a too short surface-CO distance, and a too large reduction in C-O stretch frequency. It will therefore be imperative to understand precisely in what way small clusters are deficient. In fact, a systematic investigation of the changes occurring when one goes from small clusters to extended systems, using the type of analysis introduced above, will be a most challenging task for the future.

Acknowledgement

This investigation was supported in part by the Netherlands Foundation for Chemical Research (SON) with financial aid from the Netherlands Organisation for the Advancement of Pure Research (ZWO).

REFERENCES


1. Muetterties, E.L., Rhodin, T.N., Band, E., Brucker, C.F., and Pretzer, W.R.: 1979, Chem. Rev. 79, p. 91.
2. Plummer, E.W., Salaneck, W.R., and Miller, J.S.: 1978, Phys. Rev. B. 18, p. 1673.
3. a) Elian, M. and Hoffmann, R.H.: 1975, Inorg. Chem. 14, p. 1058.
   b) Hoffmann, R.H., Schilling, B.E.R., Bau, R., Kaesz, H.D., and Mingos, D.M.P.: 1978, J. Am. Chem. Soc. 100, p. 6088. Schilling, B.E.R., and Hoffmann, R.H.: 1979, J. Am. Chem. Soc. 101, p. 3456.
4. Baerends, E.J., Ellis, D.E., and Ros, P.: a) 1973, Chem. Phys. 2, p. 41; b) 1973, Chem. Phys. 2, p. 52.
5. Baerends, E.J. and Ros, P.: 1978, J. Quant. Chem. 12S, p. 170.
6. a) Snijders, J.G., and Baerends, E.J.: 1978, Mol. Phys, 36, p. 1789.
   b) Snijders, J.G., Baerends, E.J., and Ros, P.: 1979, Mol.Phys. 38, p. 1909.
7. Kitaura, K., and Morokuma, K.: 1976, Int. J. Quant. Chem. 10, p.325.
8. Heijser, W., Baerends, E.J., and Ros, P.: 1980, J. Molec. Struct. 63, p. 109.
9. a) Ellis, D.E.: 1968, Int. J. Quant. Chem. 2 S, p. 35.
   b) Ellis, D.E., and Painter, G.S.: 1970, Phys. Rev. B2, p. 2887.
10. Baerends, E.J., and Ros, P.: 1975, Chem. Phys. 8, p. 412.
11. Ziegler, T., and Rauk, A.: 1977, Theoret. Chim. Acta 46, p. 1.
12. Ziegler, T., and Rauk, A.: 1979, Inorg. Chem. 18, p. 1755.
13. Ziegler, T., and Rauk, A.: 1979, Inorg. Chem. 18, p. 1558.
14. Bagus, P.S., Gilbert, T.L., and Roothaan, C.C.J.: 1972, J.Chem.Phys. 56, p. 5195.
15. Bachmann, C., Demuynck, J., and Veillard, A.: 1980, Faraday Symp. 14, p. 170.
16. Dunlap, B.J., Connolly, J.W.D., and Sabin, J.R.: 1979, J. Chem. Phys. 71, p. 4993.
17. Heijser, W., van Kessel, A.Th., and Baerends, E.J.: 1976, Chem. Phys. 16, p. 372.
18. Clementi, E.: 1969, Int. J. Quant. Chem. 3 S, p. 179.
19. Clementi, E., and Popkie, H.: 1972, J. Chem. Phys. 57, p. 4870.
20. Stoll, H., Pavlidou, C.M.E., and Preuss, H.: 1978, Theoret. Chim. Acta 49, p. 143.
21. Hohenberg, P., and Kohn, W.: 1964, Phys. Rev. 136 B, p. 864.
22. Kohn, W., and Sham, L.J.: 1965, Phys. Rev. 140 A, p. 1133.
23. Gunnarsson, O., and Lundquist, B.I.: 1976, Phys. Rev. B 13, p. 4274.
24. Heijser, W., Baerends, E.J., and Ros, P.: 1980, Faraday Symp.14,p.211.
25. Summerville, R.H., and Hoffmann, R.H.: 1979, J. Am. Chem. Soc. 101, p. 3821.
26. Bachmann, C., Demuynck, J., and Veillard, A.: 1980, Faraday Symp. 14, p. 170.
27. Bénard, M.: 1979, Inorg. Chem. 18, p. 2782.

# CHEMISORPTIVE PROPERTIES OF TRANSITION METAL CLUSTERS

F. Cyrot-Lackmann
Groupe des Transitions de Phases, C.N.R.S.
B.P. 166, 38042 Grenoble Cedex, France

Chemisorptive properties of simple gases on transition metal clusters are shown to be closely related to their electronic properties. These properties depend on the size and roughness of the cluster. Striking modification of the electronic properties from the bulk appear for atomic sites having a low coordination number. Spin orbit coupling play an important role for the third series, and induce differences in chemisorptive properties between Ni and Pt. Trends and order of magnitude of binding energies of simple gases, such as hydrogen, oxygen, sulphur are calculated.

## I. INTRODUCTION

The surfaces of all solids and liquids adsorb foreign molecules from the atmosphere. The adsorbed molecules change most of the chemical and physical properties of the substrate : adhesion, catalysis, corrosion, fracture... and many other important mechanisms are affected by the formation of molecular layers on a surface. Therefore a knowledge of adsorption is a necessary step for their fundamental understanding.

Adsorption on solid surfaces is a complex phenomenon as most catalytic surfaces are not uniform. They consist of sites with a broad distribution of adsorption energies and irregular surface patterns, and an important problem is the identification of active sites. Progression in the comprehension has been associated with the development of surface science, involving experimental studies on well characterized surfaces and the study of microscopic parameters, such as energy and structure associated with adsorption.

A lot of experimental data exist now on chemisorption of simple gases on transition metals and they have been recently reviewed by many authors[1,2]. Although large scattering exists in the data, some systematic trends have been observed and time is gone to construct simple theory leading to an understanding of the observed trends. One can sum up the main trends in the following way.



*R. Daudel, A. Pullman, L. Salem, and A. Viellard (eds.), Quantum Theory of Chemical Reactions, Volume III, 35–53.*

a) The heat of adsorption is of the order of the heat of formation of the corresponding compounds. The bond energy is also of the order of the dissociation energy of the corresponding molecules. Table 1 gives some numbers for the hydrogen adsorption.

| | Ni-H | Cu-H | Ag-H | Pt-H |
|---|---|---|---|---|
| $(M-A)_{mol}$ | 60 | 66 | 53 | 83 |
| $(M-A)_{chim}$ in Kcal/mole | 63 | 56 | 52 | 57 |

This is a very old finding and Langmuir in 1918 was probably the first one to express explicitely the similarities between adsorption on solids and usual chemical bonds. One can also note that the strength of the bond follows with the adsorbate the order

N > O > H > Co ≳ NO

b) There is a smooth variation of heats of adsorption with the number of d electrons of the metal : it decreases from the early to late transition metals. Fig. 1 shows for example such behaviour for oxygen and hydrogen.

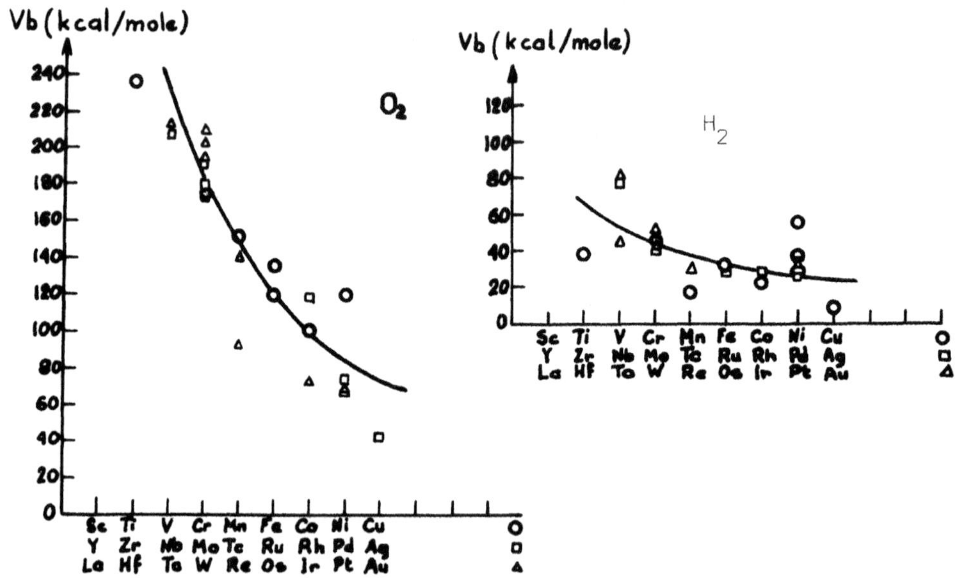


*Fig. 1. Heats of adsorption of hydrogen and oxygen on polycrystalline transition metal surfaces from (2).*

c) The heat of adsorption is not very sensitive to the crystallographic plane. One finds a change of a variation of 1 to 10 kcal/mole between various surface planes compared to 20 to 100 kcal/mole for the heat itself. No systematic trends exist for the more adsorbing plane, as there are a large scattering of the existing data. One can in some cases observe multiple binding states, particularly for stepped surfaces.

d) Interaction between adsorbates has been observed. Its leads to a decrease of the heat of adsorption when the coverage is increased. In some cases it permits the formation of ordered adsorbed phases and one can observe as a function of the temperature and coverage an order-disorder transition with typical phase diagram. The energies of interaction are of order 0.5 to 1 kcal/mole.

e) There is a high surface mobility for small adsorbed species (O, H, N, Co..). Indeed the interaction energy and the variation of heat of adsorption between different atomic sites are of the same order of magnitude as the thermal energy at room temperature ($kT \sim 0.5$ kcal). One can thus question the validity of the usual assumptions of localized adsorption and independent adsorbates on which the main approaches of the kinetics of chemisorption are widely based.

f) One generally observes a relative small charge of the work function in adsorption, leading thus to a negligible charge transfer.

g) Results of surface crystallography although tentative in some cases due to the limited data base, indicate that the adsorption sites usually maximize the number of nearest metal neighbours : the adatom chooses the hollow site. The adsorption bond lengths fall mostly within the range of known bond lengths for the same pair of atoms in molecules or bulk solids. Contraction of metal-metal bond length usually found for clusters or surfaces are also usually cancelled by adsorption.

As a conclusion, the large heat of adsorption is relatively easy to understand with the concept of the chemical bond. But the range of energy of the order of thermal energy (i.e. 0 to 1 kcal), which is crucial to understand the different behaviours of various cluster metals for chemisorption and catalysis need a precise microscopic approach on the atomic and electronic scale.

The qualitative description of chemisorption is well-known. In the process of chemisorption, when an adsorbate approaches the surface substrate, there is a broadening of its atomic levels due to the hybridization with the substrate electronic levels. Schematically, there are two limiting cases :

- for a weak adsorbate-metal interaction, the adsorbate keeps its atomic character and its atomic level is slightly broadened forming a virtual bound state,
- for a strong adsorbate-metal interaction, the system adsorbate plus its neighbouring surface atoms of the substrate behaves like a surface molecule coupled to the substrate. The atomic level of the adsorbate will

thus form two broadened states, a bonding and an antibonding one. One expects that these two regimes will give different energetics of chimisorption, the second case leading to larger binding energies. These two regimes are two extreme limits and many of the physical interesting cases are in-between.

For a more quantitative approach, two ways exist in the literature : one based on a cluster model theory, and one on the alloy analogy (absorption problem) of the adsorption based on concepts more familiar respectively in quantum chemistry or solid state physics.

The first approach pictures the metal substrate by a small cluster of metal atoms which cannot exceed 10 - 20 atoms due to numerical difficulties. This approach emphasizes the role of the surface molecule picture and treats the electrons in a localized scheme. One thus immediately obtains a bond energy of the order of magnitude of the dissociation energy of the corresponding molecule. However, detailed results are not reliable, as such small clusters cannot simulate real solid surfaces. For instance, the ionization potential and the electron affinity of such clusters are different and are also different from the work function of bulk solids. Similarly, the bandwidth, the cohesive energy are different from the bulk or from large clusters. For example, we calculate that for a cubooctahedron of 55 atoms, the cohesive energy is 80 % of the bulk and 87 % for a cluster of 147 atoms, i.e. within 0.5 to 1 kcal. Difficulties are also encountered in this model to treat the interaction between adsorbates and all problems related to coverage.

The second approach describes the metal substrate through a band structure model theory with itinerant electrons similar to that used in bulk solids. It uses similar techniques to those developed for treating impurities in metals. The two situations are very similar and many known results of solid state physics can be translated for chemisorption. The strength of the interaction between the adsorbate and the substrate will therefore not only depend on the surface molecule directly formed when adsorbing but also on the whole band structure of the metal due to an indirect interaction through the rest of the substrate. In such an approach, the order of magnitude of the chemisorptive bond is not so obvious to predict. However, we will show in the following that in a tight binding model, we recover the order of magnitude of the dissociation of molecules. But the interest of this approach stems from the possibility to handle problems directly related to the detail atomic description of the solid surface and which are outside the possibilities of the cluster model theory such as the size effect ; the heterogeneity of surfaces ; possible magnetic effects ; the importance of spin orbit coupling which enables to understand differences between Ni, Pd, Pt ; the effects of coverage.

## II. ELECTRONIC STRUCTURE OF TRANSITION METALS : TIGHT BINDING SCHEME

The band electronic structure of a transition metal is formed by a narrow d band with a high density of states overlapping with a broad sp band. Many of the peculiar properties of transition metals have been shown to be related to the d character of their valence states and the sd hybridization can be neglected as a first approximation. The smallness of the overlap between the d atomic orbitals on neighbouring atoms justifies the use of the simple LCAO molecular orbital theory (the tight binding scheme analog to the Huckel one) to describe the d band of transition metals[3].

In the LCAO method, the wave function $\Phi$ of an electron is expressed in terms of linear combinations of atomic orbitals $\phi_\lambda(\underset{\sim}{r}-\underset{\sim}{R}_i)$ :

$$H \Phi = E \Phi$$

$$\Phi = \sum_{i,\lambda} A_{i\lambda} \phi_\lambda(\underset{\sim}{r}-\underset{\sim}{R}_i) = \sum_{i,\lambda} A_{i\lambda} \phi_{\lambda i} \tag{1}$$

where i labels the site and $\lambda$ the orbital degeneracy

$$H = T + \sum_i V(\underset{\sim}{r}-\underset{\sim}{R}_i) = T + \sum_i V_i \tag{2}$$

$$(T+V_i)\, \phi_{\lambda i} = E_{0\lambda}\, \phi_{\lambda i} \tag{3}$$

$E_{0\lambda}$ being the atomic orbital level.

Eqs. (1,2,3) applied to the Schroedinger equation lead to the secular determinant

$$\det \left| H_{ij} - E\, S_{ij} \right| = 0 \tag{4}$$

for the energy values E, where $S_{ij}$ and $H_{ij}$ are the overlap and hamiltonian matrices

$$S_{ij} = \langle \phi_i \mid \phi_j \rangle$$

$$H_{ij} = \langle \phi_i \mid H \mid \phi_j \rangle$$

In general orbitals on different sites are not mutually orthogonal, so that $S_{ij} \neq 0$ for $i \neq j$, although we assure the orbitals normalized, i.e. $S_i = 1$. But the overlap is usually small and can be neglected as a first approximation. $H_{ij}$ can be expressed in terms of two center integrals, respectively of the type

$$\langle \phi_i \mid V_i \mid \phi_j \rangle = t_{ij} \tag{5}$$

called hopping (or resonance) integrals and leading to the width of the band ($W \sim zt$, z = number of coordination) and $\langle \phi_i \mid V_j \mid \phi_i \rangle = \alpha_i$ called crystalline field (or Coulomb) integrals leading to the shift of the band respective to the atomic level $E_0$.

Several recent works permit to put the tight binding method onto a proper quantitative basis. These works have been guided by the success of the semiempirical fitting or interpolation scheme. If some information is known from experiment or accurate ab initio computation about the energy levels of a molecule or a crystal, e.g. energies at points of high symmetry in the Brillouin zone, then one can fit a few parameters in (4) and hence interpolate the remainder of the energy bands. This is the way mostly used to get the tight binding parameters. This success has thus raised theoretical questions, and recent works throw light on the justification of tight binding scheme by discussing the effect of the overlap between the atomic orbitals, the suitability of atomic orbitals as basis function and the treatment of electron correlation. This leads to the understanding of the physical ideas of the tight binding scheme. The $H_{ij}$ are interatomic quantities but are given by an atomic description in terms of intraatomic quantities characteristic of the nature of the free atom. The theory defines thus a standard transition metal with a canonical shape for band structure which can be used for scaling between different elements, crystal structures and interatomic spacings.[3,4,23]

The theory can be extended to alloys and to compounds with s or p bands of tight binding form. One can for example apply parameters from the metal-oxygen bond to oxides or to chemisorption of oxygen on metals. The parameters are thus transferable quantities which permit to transfer information in various situations. One recovers thus the idea of the chemical bond associated with delocalized (or itinerant) electrons.

Finally, let us point out that while in a periodic solid the use of Bloch's theorem allows simplification of Eq. (4), for an arbitrary cluster, the size of the matrices H and S equals the total number of orbitals. For small clusters (up to some tens of atoms), one can diagonalize directly, but for larger clusters or solids with rough surfaces, one cannot calculate the eigenstates. We will use thus a moments method associated with a continued fraction expansion so as to calculate directly the electronic density of states without having to know the eigenstates.[5]

## III. ELECTRONIC PROPERTIES OF TRANSITION METAL CLUSTERS AND SURFACES

The study of the change of the electronic properties of a transition metal when going from the bulk to semi-infinite crystals and to clusters of various size and geometry is of fundamental importance for the understanding of the mechanism of chemisorption. The following questions are of particular interest :

- at what size does a cluster behave as a bulk or a semi-infinite surface,

- what are the essential features of the electronic properties of active sites for adsorption,

- what are the stable geometry of clusters, and how does this change with adsorption.

A useful tool to study the electronic properties of a peculiar atomic site $R_i$ is the local density of states (L.D.S.) $n_i(E)$ on the site $R_i$

$$n_i(E) = \sum_{\lambda,n} \left[A_{i\lambda}(E_n)\right]^2 \delta(E-E_n)$$

where $E_n$ are the eigenstates of the system and the $A_{i\lambda}$ are defined by Eq. (1).

The L.D.S. are sensitive to the localization of the electrons near the site $R_i$ and give information on the spatial distribution of electrons in the various orbitals $\lambda$.

III.1 Semi-infinite transition metal surfaces

One can resume the detailed results obtained for various cleavage pleanes and various crystallographic structures in the following way[6]

- For a dense cleavage plane (i.e. (111) for an FCC lattice, (110) for a BCC one), the L.D.S. on the surface behaves like the bulk DOS (Fig.2a).

- For a non dense cleavage plane such as the (110) for a FCC lattice or the (100) for a BCC lattice, it exists a virtual bound state peak in the middle of the band. This peak is due to the resonance of the surface atoms interacting weakly with the bulk ones[7]. There is correlatively a weakening in the shape of the L.D.S. at the top of the band (Fig. 2b).

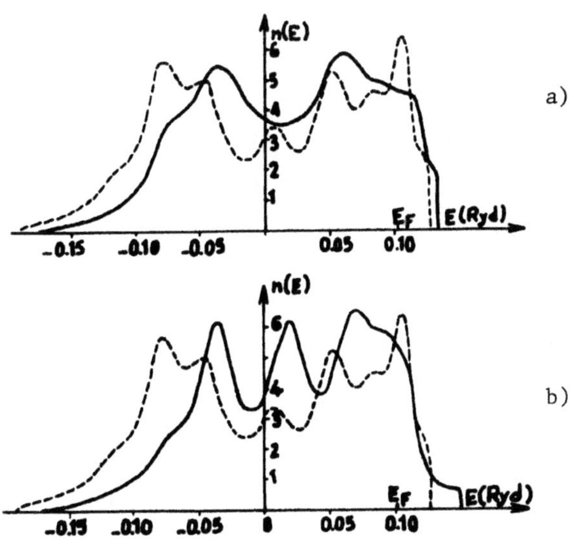


*Fig. 2. Local density of states on the surface of nickel compared with the bulk one (broken curve) a) (111) plane ; b) (100) plane.*

- When one enters in the crystal, the L.D.S. tend to have a behaviour similar to the bulk one.

These features of the L.D.S. on surfaces exist for all the transition metals and have been observed by spectroscopy. They lead to a slight increase in the asphericity of the charge density of the d electrons when going from a bulk to a surface site giving indications on the possible adsorption sites[8].

III.2 Stepped surfaces

Sometimes dramatic differences of reactivity between flat surfaces and stepped surfaces (i.e. high index surfaces) were reported, particularly in the case of Pt[9]. This suggests that electronic properties may differ markedly from surface site to surface site in heteregeneous surface. This is indeed what we have found in a calculation concerning L.D.S. on various sites of stepped Ni and Pt[10,11]

A sharp surface virtual bound state peak is found in the L.D.S. at the protruding edge of the stepped surfaces of Pt and the symmetry of states near the Fermi level is found to be rather dependent on the geometry of the surface[10] (Fig. 3).

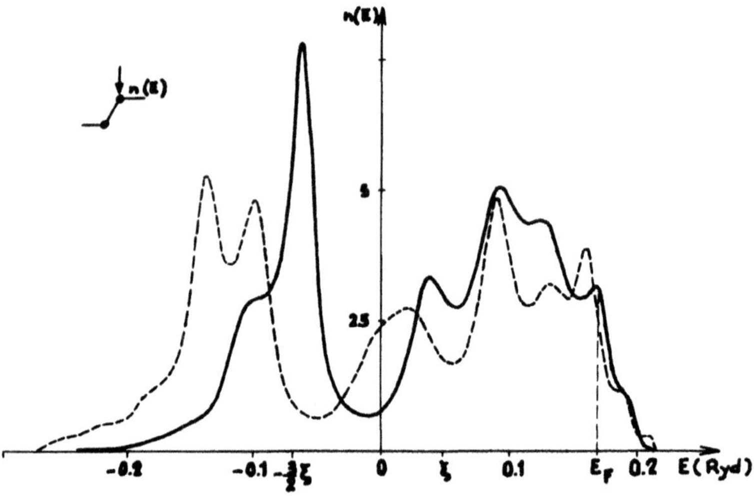


*Fig. 3. The L.D.S. on the first atomic ion of stepped Pt |6(111)x(100)| compared to the bulk one (broken curve).*

In the case of Ni, the conclusions are rather different and stepped surfaçes behave like the non dense low index planes[11]. Spin orbit coupling $\xi.\vec{L}.\vec{S}$. play thus an important role in the case of Pt (where $\xi$ = 0.42 eV) in contrast to the case of Ni ($\xi$ = 0.07 eV) where it is neglectible.

III.3 Clusters

To ensure a maximum surface area effect, catalysts are clusters usually made of powders supported on an inert holder and a major problem in heterogeneous catalysis is characterization of the active sites. A way to get solution, is to find a correlation between particle size and some catalytic properties, selectivity or specific rates. It is thus of particular interest to study the electronic factors associated with change in the size and geometry of clusters.

We have studied Ni and Pt clusters of size varying between 13 atoms to 2089 atoms (i.e. dispersion rate of 0.92 to 0.30 and a diameter of 5 to 40 Å for Ni) and of different geometries of cubic and five fold symmetry, such as icosahedron and cubooctahedron[11,12,13]. Experimentally one observes transition metal clusters of five fold symmetry, these clusters making a transition towards an fcc lattice when the size is increased[14].

Our results for the L.D.S. described in detail in refs.[11,12,13] show the following trends :

- the band features of clusters approach those of bulk for large size, i.e. clusters having a few hundreds atoms

- it appears extra peaks of surface resonant states in the middle of the band in some atomic surface sites having a low coordination number. These features are similar to those existing in the case of rough surfaces, but enhanced by the size effect. In particular, spin orbit coupling has a spectacular effect in the case of Pt clusters, as shown in Fig. 4. The L.D.S. on the corner atoms show indeed in that case the opening of a pseudo-gap in the band, contrary to the case of Ni clusters where there exists a central peak (Fig. 4). This has interesting consequences on the chemisorptive properties of these clusters, and is currently studied.

We have also studied the relative stability of these clusters by minimizing their total energy as a function of the interatomic distance. Results show that there is a contraction of the bond lengths larger for smaller clusters size and larger for icosahedron than for cubooctahedron. For example, for a 15 Å Ni cluster, the contraction is of the order of 3 % for fcc and 5 % for icosahedron. These results are in agreement with recent experimental work on Ni and Pt catalysts by EXAFS technique[15]. The isosahedra structure is slightly favoured for small clusters size with respect to fcc, up to Ni clusters of 15 Å but difference of energies between both structures is small, i.e. $\Delta E \sim 500$ cal/mole so that they can both exist (Fig. 5).

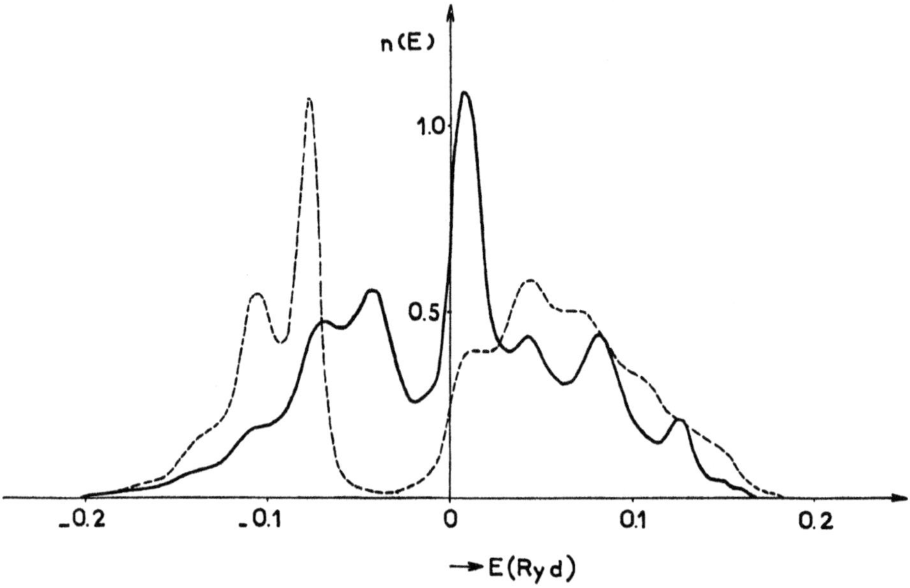


Fig. 4. Comparison of LDS on the corner atom of the cubooctahedra of 309 atoms in the absence (——) and presence (---) of spin orbit coupling.

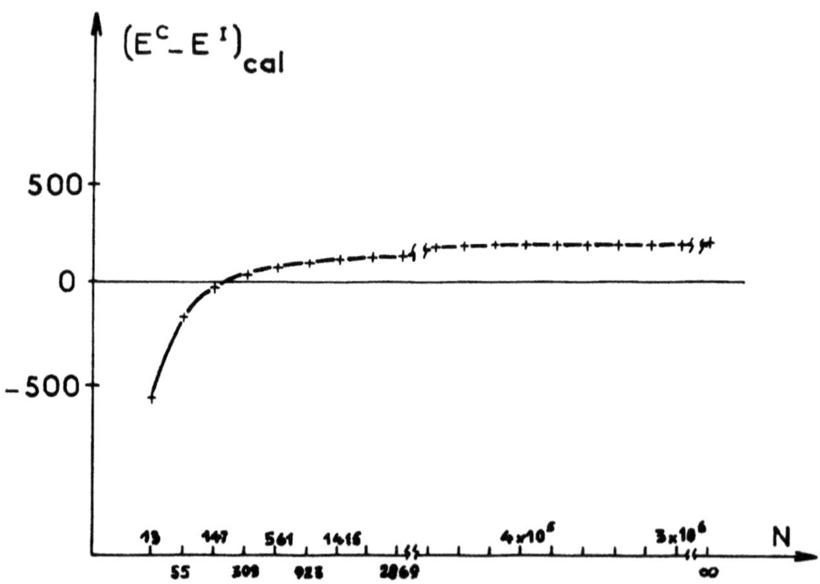


Fig. 5. Difference of cohesive energies of cubooctahedral and icosahedral Ni clusters as a function of size.

## IV. CHEMISORPTION

Many attempts have been made to correlate the electronic factors of transition metals, such as their empty d orbitals, with their power to chemisorb, including for example the application of Pauling's valence bond theory or more recent quantum mechanical treatments of clusters or surfaces. But these approaches have some serious drawbacks, being generally qualitative models representing very roughly the complexity of the surfaces and of the chemisorptive bond.

However, any attempt at an understanding of adsorption must take into account the reality of the electronic structure of the surface of the catalyst. The important changes occurring in the electronic properties when going from the bulk to surfaces or to clusters, described in the above section will affect the chemisorption process.

### IV.1 General description of the chemisorptive bond

When an adsorbate approaches a metal surface, there is an hybridization of its discrete electronic levels with the band of the metal leading to a broadening in virtual or resonant states. To describe the adsorption of an atom A on a substrate, one can write the total hamiltonian as a sum of three terms

$$H = H_a + H_m + H_c$$

where $H_a$ describes the free adsorbate characterized by an atomic level $\varepsilon_a$ with $z_a$ electrons, $H_m$ the substrate characterized by its band structure, i.e. the L.D.S. at the surface $n_s(E)$ and $H_c$ the coupling between both. This coupling can be characterized by the effective overlap B between the adsorbate and the $Z_s$ corresponding atomic sites of the substrate and by the charge transfer occurring between the adsorbate and the substrate leading to a change in the atomic level $\varepsilon_a \rightarrow \varepsilon_a^*$ of the adsorbate.

The usual approaches are based on a one electron scheme, either a tight binding model or a Hartree Fock model hamiltonian of Anderson-Friedel type. These two approaches are very similar and differ only in the way of treating the charge transfer, i.e. of calculating $\varepsilon_a^*$. In the tight binding model that we are using in the following of this paper, the charge transfer is calculated through the Friedel sum rule : the charge $z_a$ brought by the adsorbate is equal to the charge variation induced on the substrate and the adsorbate when coupled. In fact, we are using a separate charge neutrality condition for the adsorbate and the substrate as experimentally the change in work function indicates a neglectible charge transfer ($\simeq$ 0.1 $e^-$)[1]. In the Hartree Fock treatment, the change in the position of the adsorbate level is due to the Coulomb repulsion U between two electrons, i.e. $\varepsilon_a^* = \varepsilon_a + U \langle n \rangle$ where n is the number of electrons. But the values one can deduce for the charge transfer are quite important ($\simeq$ 1-2 $e^-$) and this approach seems to suffer from serious and unrealistic drawbacks.

The L.D.S. on the adsorbate is given by an equation of similar type for both approaches

$$n_a(E) = \frac{\Gamma(E)}{\left[E - \varepsilon_a^* - \Omega(E)\right]^2 + \Gamma^2(E)} \quad . \qquad (6)$$

with $\Omega(E) = \frac{1}{\pi} P \int \frac{\Gamma(E')}{E-E'} dE'$

$\Gamma(E) = \pi Z_s B^2 n_s(E)$ is the chemisorption function*. One finds easily on this expression for $n_a(E)$ the two limiting cases, virtual bound state regime or surface molecule regime for respectively a weak or a strong coupling, i.e. depending on the strength of $\Gamma(E)$.

The strength of the coupling will depend obviously on the effective overlap between the adsorbate and the corresponding substrate site, on its coordination number and also on its local density of states. This last term will be particularly important when there exist resonant surface states close to the Fermi level.

IV.2 Adsorption and magnetism

The understanding of surface magnetic properties of transition metals or their change when they participate in an adsorptive process still remains a challenging problem. However, magnetic techniques are often used to characterize the state of dispersion of supported metal catalysts, and to throw some light on the intermediate states produced in the adsorption process[16,17]. Any model description of the adsorption process must then permit to explain these magnetic properties.

When adsorbing a gas such as $H_2$, $O_2$, CO.. on a nickel substrate, there will be a braodening of the adsorbate electronic levels and a displacement of the relative positions of the s and d bands of the nickel. The very small charge transfer between the adsorbate and the substrage is shown by the small change in work function when adsorbing[1,18]. These shifts of the s and d bands will cause a reduction of the local density of states at the Fermi level on the active sites of nickel for adsorption and on its first neighbours.

These conclusions are confirmed for example by a calculation for oxygen overlayers on Ni (100)[19]. Experimentally oxygen atoms are found to be in fourfold coordinated sites, forming p(2x2) or c(2x2) overlayer. Their distance to the substrate is rather well determined, and various spectroscopic investigations have also been performed. Our results show very similar trends in the change of the electronic structure of Ni (100) when adsorbing oxygen with various converages (Fig. 6). There is a shift of the d band towards the negative energies and consequently a drastic reduction of the L.D.S. at the Fermi level. A simple

---

* $n_s(E)$ represents the LDS at the surface only for simple adsorbate geometries, otherwise one has to consider the orbital group density of states

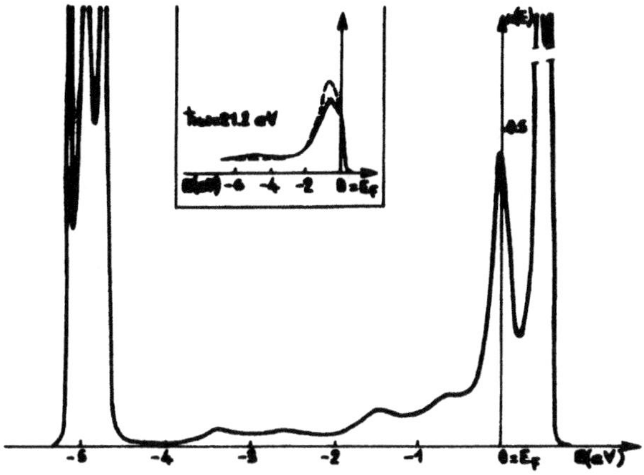


*Fig. 6a.* LDS on the p(2x2) oxygen overlayer on Ni(100).
In the insert, the photoemission spectra |21| ($\hbar\omega$ = 21.2 eV) ;
(---) clean surface, (——) surface with O overlayer.

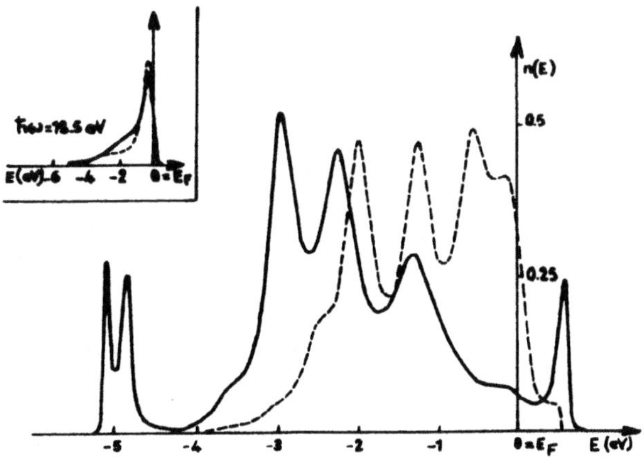


*Fig. 6b.* LDS (——) on the first Ni(100) plane with an adsorbed p(2x2) oxygen overlayer compared with the clean surface (---). In the insert, the photoemission spectra |21| ($\hbar\omega$ = 13.5 eV) ; (---) clean surface, (——) surface with O overlayer.

reasoning using the Stoner criterion relating the existence of ferromagnetism to a high value of the density of states at the Fermi level will thus lead to a demagnetization of the active sites of nickel. The results for a very low coverage ($\theta \to 0$) indicate thus that one atom of oxygen demagnetizes 4 atoms of nickel. For a p(2x2) oxygen overlayer ($\theta$ = 1/4) the surface layer of nickel is likely to become non-magnetic. This is in agreement with experiments which have revealed a decrease of the saturation magnetization of Ni when adsorbing oxygen[20].

Another case of interest is the correlative change in the magnetic properties of nickel and nickel-copper catalysts when chemisorbing hydrogen, oxygen and in the binding energies which have been shown to depend on the structure of the catalyst[17,20,22]. This effect can be explained as due to the difference in the electronic properties of surface nickel atoms with the roughness of surface or alloy composition.

IV.3 Binding energy

In a tight binding scheme, the binding energy is given by

$$U(M\text{-}A) = \int^{E_F} E\,\delta n(E)\,dE \tag{7}$$

where $\delta n(E)$ is the change in the density of states of the system due to the adsorption of adatom A on metal M and $E_F$ is the Fermi level. Let us remind that the binding energy can be deduced from the heat of adsorption $\Delta H_{ads}$ through

$$U(M\text{-}A) = \Delta H_{ads}$$

$$\text{or} \quad U(M\text{-}A) = \Delta H_{ads} + \frac{D(A_2)}{2}$$

for the case of atomic and molecular adsorptions respectively, where D is the dissociation energy of the $A_2$ gas molecule. $\delta n(E)$ will be composed on two parts : the first one is the local density of states (LDS) on the adsorbate, given in a one electron scheme by Equ. (6) ; the second one is the change of the LDS on the atoms of the metal. We can reasonably assume that this change is non negligible only on the atoms of the substrate directly linked to the adsorbate and on their nearest neighbours.

The binding energy is evaluated through Eq. (7) using the one electron tight binding scheme described in Section IV.1. In particular we assume that there is no charge transfer between the adatom and the metal. This stems from the small change of work function during adsorption[1]. We have seen that the strength of the coupling depends on three factors : the adsorption site geometry, the LDS on the substrate adsorption sites and the direct coupling B between the electronic states of the adsorbate and the substrate. This last term writes as a function of the hopping integrals

$$\begin{aligned} \langle \phi_A | V_A | \phi_M \rangle &= t_{AM} \\ \langle \phi_M | V_M | \phi_A \rangle &= t_{MA} \end{aligned} \tag{8}$$

where $\phi_A$ and $\phi_M$ are respectively the atomic orbitals on the adsorbate and the metal atoms directly linked, and $V_A$ and $V_M$ the corresponding potentials. Let us remind here that the LDS on the substrate will be given in terms of the $t_{MM}$ hopping integrals between two metal atoms (see §2 and Eq. 5). These hopping integrals between orbitals on adjacent atoms are usually defined in terms of Slater-Koster formulas[24] writing the s, p,d orbitals as a linear combination of σ, π, δ functions refering to the component of angular momentum around the axis joining the atoms. Thus, for example for a transition metal, the $t_{MM}$ are defined by three parameters ddσ, ddπ, ddδ. For the coupling between hydrogen and a transition metal, $t_{HM}$ is given by one parameter sdσ and for the coupling between oxygen (or sulphur) and a transition metal, $t_{OM}$ (or $t_{SM}$) is given by two parameters pdσ and pdπ.

In order to describe the main trends of variation of the binding energy, we will use a simplified description of the LDS through its first few moments and a continued fraction expansion. We introduce the exact second moment in the first step of the continued fraction and sum up then the fraction[25,26]. The second moment of the LDS on the adsorbate writes as

$$\begin{aligned} b_a &= Z_s (sd\sigma)^2 \quad \text{for hydrogen} \\ b_a &= Z_s \frac{1}{3} \left[ (pd\sigma)^2 + 2(pd\pi)^2 \right] \quad \text{for oxygen or sulphur} \end{aligned} \tag{9}$$

where $Z_s$ is the number of coordination of the adsorption site ($Z_s$ = 1 for an adsorption site ; $Z_s$ = 2 for bridge site ; $Z_s$ = 3 or 4 for centred site). The second moment of the LDS on the substrate writes as

$$b = 12 \frac{1}{5} \left[ (dd\sigma)^2 + 2(dd\pi)^2 + 2(dd\delta)^2 \right] \tag{10}$$

The first contribution U to the binding energy comes from the broadening of the adsorbate levels which is given in this model by :

$$n_a(E) = \frac{b_a}{2\pi} \frac{(4b-E^2)^{1/2}}{b_a^2 + (b-b_a)E^2} \quad \text{if } b_a < 2b$$

If $b_a > 2b$, i.e. if the coupling is strong, we also get two bound states which emerge from the band at energy $\pm\, b_a/(b_a-b)^{1/2}$ with a weight $(b_a-2b)/2(b_a-b)$*. In the usual cases of adsorption, we are generally in the strong coupling limit (see Table II). One gets then for the first contribution $U_1$ to the binding energy for example in the case of hydrogen adsorption

$$U_1 = -2 \left[ \frac{b_a(b_a-2b)}{2(b_a-b)^{3/2}} + \frac{b_a}{2\pi(b_a-b)} \left( 2\sqrt{b_a} - \frac{b_a-2b}{\sqrt{b_a-b}} \operatorname{arctg} 2\sqrt{b_a} \frac{\sqrt{b_a-b}}{b_a-2b} \right) \right] \tag{11}$$

For oxygen or sulphur adsorption, the formula is more complicated due to the partial filling of the adsorbate p band[26].

---

* Let us remark here that this simple model permits to recover splits off states well-known to appear in the strong coupling limit[1,23,27].

The second contribution $U_2$ to the binding energy comes from the adsorption sites of the metal. It can be simply evaluated by the change of the second moment of surface atoms. From the formula relating the change $\delta b_2$ of second moment $b_2$ of the density of states to a change $\delta E_c$ in the cohesive energy[5]

$$\frac{\delta E_c}{E_c} = \frac{1}{2}\frac{\delta b_2}{b_2}$$

We thus get for the change of energy of the adsorption site

$$U_2 = \frac{1}{2} Z_s \frac{3b_a}{20b} E_c' \qquad (12)$$

where $E'_c$ is the cohesive energy of a surface atom.

A calculation of the binding energy needs the knowledge of the second moments $b_a$ and $b$ given by Eqs. (9) and (10) ; it is thus necessary to evaluate the hopping parameters $t_{AM}$ as a function of the adsorption band length. Up to now, no such quantities have been yet computed and the data used were obtained by a fit to the corresponding bulk compounds band structure calculation, when these exist through an interpolation scheme. This was the way we proceed for the calculation of the electronic spectra of oxygen adsorbed on Ni (110)[19] (Fig. 6). But this is not satisfying especially for energy calculations, as one has an estimation of these parameters for only a fixed distance A-M corresponding to the bulk compound. We have thus done a calculation[28] of these hopping parameters from first principles using the atomic structure calculations of Herman and Skillman[29] done through a Hartree-Fock-Slater approach. The detailed results are described in ref.28. A typical curve giving the variation of these parameters as a function of the distance between A and M is given in Fig. 7 for Ni-H. Some results are shown in Table II for adsorption on Ni for a coordination $Z_s = 3$ of adsorption site.

For metals at the end of the series, such as Ni, Pd, Pt, due to their small cohesive energy, the order of magnitude of the binding energy is given by the first contribution $U_1$ (see Table II). This term varies slightly when going from the beginning of the series to the end. But the variation of the binding energy along the series is dominated by the second term $U_2$, mainly due to the large variation of parabolic shape of the cohesive energy with the series[3]. Moreover the hopping parameters $t_{AM}$ and $t_{MN}$ decreases with the filling of the series[28]. Both effects will lead to a decrease of the binding energy along the series, in agreement with experimental results particularly for oxygen or sulphur adsorption. The variation of the binding energy of hydrogen within the series is smoother than for the other simple gases, feature existing also for the heat of formation of hydrids[2]. Such a behaviour cannot be accounted by this simple model and could be attributed to the importance of correlation effects in the case of hydrogen[30].

Many other factors can affect the magnitude of the binding energy. One of them is the LDS on the adsorption site, particularly when exist

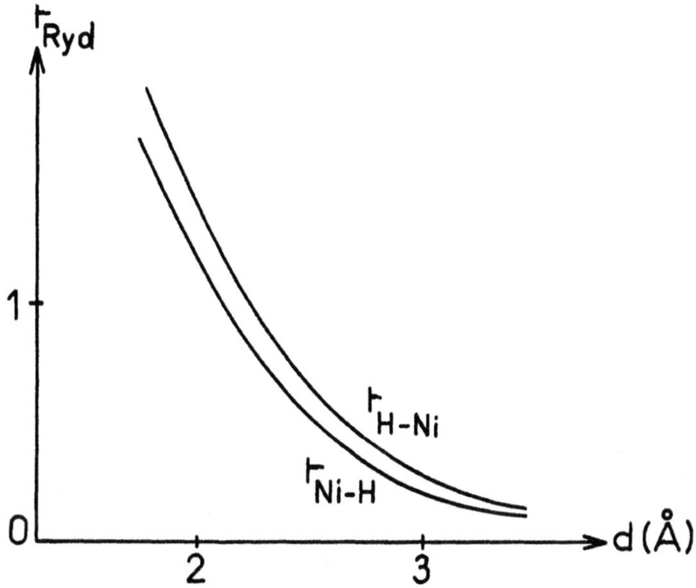


*Fig. 7.* Hopping integrals for a Ni-H bond as a function of Ni-H distance.

TABLE 2

| | d(Å) | $b_{a(eV)}^2$ | $U_{1eV}$ | $U_{2eV}$ | $U_{tot\,eV}$ | $U_{expl\,eV}$ |
|---|---|---|---|---|---|---|
| Ni-H | 1.86 | 12 | 3 | 0.5 | 3.4 | 2.8 |
| Ni-O | 2.08 | 4.37 | 3.6 | 0.7 | 4.2 | 4.3 |
| Ni-S | 2.28 | 4.64 | 4.5 | 0.7 | 5.2 | 4.5 |

with b equal to 1,54(eV)$^2$ for pure Ni with an interatomic distance of 2-46 Å.

resonant surface states close to the Fermi level[18,23,32]. In that case, one can find examples where one goes from a weak coupling regime to a strong coupling regime when adsorbing respectively on a dense plane and a rough plane of a transition metal of the middle of the series. The surface resonances are thus quenched by the effect of metallic adsorbates[18,31]. This will increase the anisotropy of the binding energy. This may be an explanation of the large discrepancies one observe in the experimental résults of heats of adsorption of simple gases on single crystal surfaces of transition metals of the middle of the series[2]. The spin orbit coupling will also play an important role as shown in § III for metals of the third series such as Pt and illustrated by the peculiar effects of steps[1,9]. Experiments show that there can be many binding states of an adsorbed small molecule on a given surface indicating that even a single crystal plane may be heterogeneous when viewed by the adsorbed species. This effect is enhanced in the case of adsorption on small aggregates such as those used in catalysis. In that case, the way of preparation of the catalyst or the metal particle size induce important charges in the catalytic reactions which can be related to difference in electronic properties with the surface structure[17,20,22,23,32].

Another problem in the calculation of the binding energy stems from the large variation of the hopping parameters $t_{AM}$ with the adsorbate-metal distance (see for example Fig. 7 and Ref. 28) and the uncertainty on the band adsorption length makes difficult a precise determination of the binding energy. The binding energy should be minimized as a function of the distance to determine the adsorption geometry, in a way similar to what we have done for studying the relaxation and stability of nickel clusters[11,12] (see III.3), taking into account also the repulsive part of the energy. But one has to be aware that a full calculation would need a self-consistent determination of both the adsorption geometry and the electronic configuration, calculation which has yet never been done for surfaces or clusters larger than a few atoms. Indeed, the electronic configuration depends on the relative positions of the s and d bands and will change in surface or cluster from the bulk tending to an atomic configuration for very small clusters and will also change when adsorbing a gas.

Acknowledgements

It is a pleasure to thank Drs. Y. Boudeville, M. Cyrot, J. Friedel, P.C. Gravelle, B. Imelik, S.N. Khanna and G. Martin for useful discussions.

References

1. Somorjai G.A., *Principles of Surface Chemistry*, Prentice Hall, 1972.
   Rhodin T.H. and Ertl G., *The Nature of the Surface Chemical Bond*, North Holland, 1979.
   Smith J.R., *Theory of Chemisorption*, Springer Verlag, 1980.
   Oudar J., Catal. Rev. 22, 1980, pp. 171.
2. Toyoshima I. and Somorjai G.A., Catal. Rev. 19, 1979, pp. 105.

3. Friedel J., *The Physics of Metals*, Ed. Ziman J.M., 1969.
Heine V., Solid State Physics 35, 1980, pp. 1
4. Anderson P.W., Phys. Rev. Lett. 21, 1968, pp. 13.
Fricker H.S. and Anderson P.W., J. Chem. Phys. 55, 1971, pp. 5028.
Cyrot-Lackmann F. and Del Re G., to be published.
5. Cyrot-Lackmann F., J. de Physique C1, 1970, pp. 67.
Gaspard J.P. and Cyrot-Lackmann F., J. Phys. C6, 1973, pp. 3077.
6. Desjonquères M.C. and Cyrot-Lackmann F., J. Phys. F 5, 1975, pp. 1368.
Desjonquères M.C., Solid St. Comm. 20, 1976, pp. 855.
7. Friedel J., J. de Physique 37, 1976, pp. 883.
8. Desjonquères M.C. and Cyrot-Lackmann F., J. Chem. Phys. 64, 1976, pp. 3707.
9. Somorjai G.A., Adv. in Catal. 26, 1977, pp. 1.
10. Desjonquères M.C. and Cyrot-Lackmann F., Solid St. Comm. 18, 1976, pp. 1127.
11. Gordon M.B., Thèse de 3e Cycle, Grenoble, 1978.
12. Gordon M.B., Cyrot-Lackmann F., Desjonquères M.C., Surf. Sci. 80, 1979, pp. 159.
13. Khanna S.N., Cyrot-Lackmann F., Boudeville Y., Rousseau J., Surf. Sci., to be published.
14. Renou A. and Gillet M., Thin Solid Films 41, 1977, pp. 15.
15. Renouprez A., Fouilloux P. and Moraweck B., Studies in Surf. Sci. Catal. 4, 1980, pp. 421.
Apai G., Hamilton J.F., Stohr J. and Thompson A., Phys. Rev. Lett. 43, 1979, pp. 165.
16. Selwood P.W., *Chemisorption and Magnetization*, Academic Press, 1975.
17. Derouane E.G., Simoens A., Colin C., Martin G.A., Dalmon J.A. and Vedrine J.C., J. Catal. 52, 1978, pp. 50.
Martin G.A., Dalmon J.A. and Dutartre R., Studies in Surf. Sci. Catal. 4, 1980, 467.
18. Christmann K., Schober O., Ertl G. and Neumann M., J. Chem. Phys. 11 1974, pp. 4528.
19. Desjonquères M.C. and Cyrot-Lackmann F., Surf. Sci. 80, 1979, pp.208
20. Dalmon J.A., Martin G.A. and Imelik B., "Thermochimie", Ed. CNRS, n°201, 1971, pp. 593.
21. Nguyen T.T.A., Cinti R.C. and Avignon M., Proc. 3rd Int. Conf. on Solid Surfaces, Vienna 1977, pp. 493.
22. Prinsloo J.J. and Gravelle P.C., J.C.S. Faraday 76, 1980, 2221.
23. Cyrot-Lackmann F., Studies in Surf. Sci. Catal. 4, 1980, pp. 241.
24. Slater J.C. and Koster G.F., Phys. Rev. 94, 1954, pp. 1498.
25. Cyrot-Lackmann F., Phys. Rev. B22, 1980, pp. 2744.
26. Cyrot-Lackmann F., to be published.
27. Cyrot-Lackmann F., Desjonquères M.C. and Gaspard J.P., J. Phys. C7, 1974, pp. 925.
28. Boudeville Y., Rousseau J., Cyrot-Lackmann F. and Khanna S.N., to be published.
29. Herman F. and Skillman S., *Atomic Structure Calculations*, Prentice Hall, 1963.
30. Friedel J., private communication.
31. Richter L. and Gomer R., Surf. Sci. 83, 1979, pp. 93.
32. Gault F.,Garin F. and Maire G.,Studies Surf. Sci. Catal. 4,1980, pp. 451

# GAS, ORGANIC SOLID STATE REACTIONS AND THEIR APPLICATIONS

Guy BERTHOLON

Laboratoire de Chimie Industrielle, Groupe de Recherches sur les Phénols, associé au CNRS (ERA 600), Université Claude Bernard Lyon I, 43, bd du 11 novembre 1918 - 69622 VILLEURBANNE CEDEX - FRANCE

In this paper the author prove that the organic solid state is a reactive one. Catalysed gas-solid reactions such as hydrogenation or uncatalysed reactions as halogenation have been chosen to illustrate this reactivity. Using single crystals, the knowledge of crystal structure and morphology allow the explanation of the observed reactivity and selectivity. The crystal chirality leads to an absolute assymetric synthesis using achiral reactants and catalyst. Theoretical chemistry can help in understanding the mechanisms of such reactions.

## I. INTRODUCTION

In the field of organic chemistry most of the synthesis reactions are performed using the reactants in the fluid state i.e. gaseous, melt or dissolved state. On the contrary in the inorganic chemistry the gas-solid reactions are well known. Why so few reactions of organic chemistry are carried out using one or more reactants in the solid state ? Two kinds of arguments are available. First, many people do not believe that the organic solid state is an unreactive one. The purpose of this paper is to show that this sentence is not really. Secondly the chemists have developped many suitable solvents, which make possible to bring into contact any reactants in the dissolved state. In the eightenth century, in the absence of suitable solvents any organic reactions were performed, in the solid state.

Furthermore in this chemistry, the explanation of results requires the knowledge of crystal structure and the crystal structure determinations of molecular crystals have made great progress in the last years. G.M.G SCHMIDT who was at the same time chemist and cristallographer, proposed the scientific ground of the organic solid state chemistry.

These two abilities had allowed G.M.J. SCHMIDT and H. MORAWETZ to put forward the topochemical postulate and the criterion of solid state reactivity (1, 2). The main work of G.M.J. SCHMIDT and after, from the Weizmann Institute of Science at Rehovot (Israël) concern essentially



*R. Daudel, A. Pullman, L. Salem, and A. Viellard (eds.), Quantum Theory of Chemical Reactions, Volume III, 55–75.*

**CARBOXYLATION**

Kolbe and Schmitt

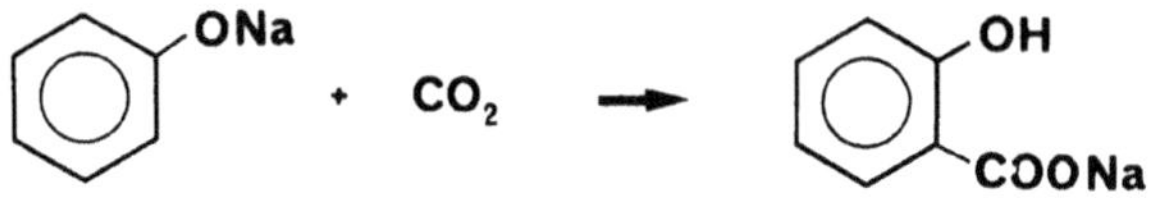


**DEHYDROGENATION**

Ressler

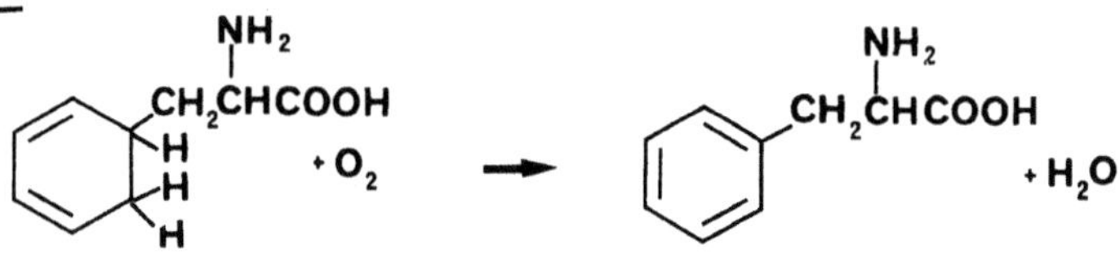


**OXYDATION**

Hochstrasser

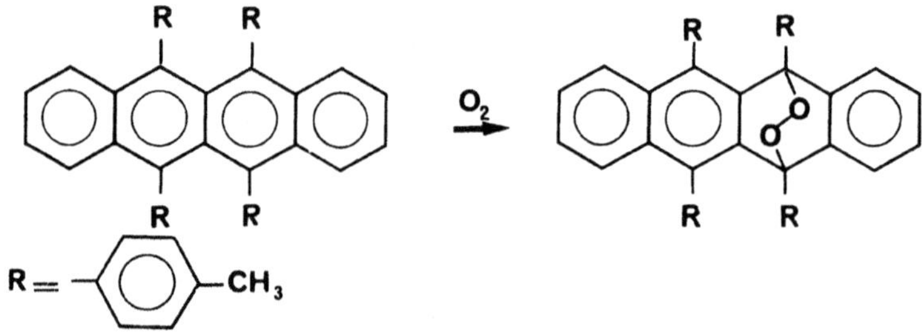


**OZONOLYSIS**

Desvergne and Thomas

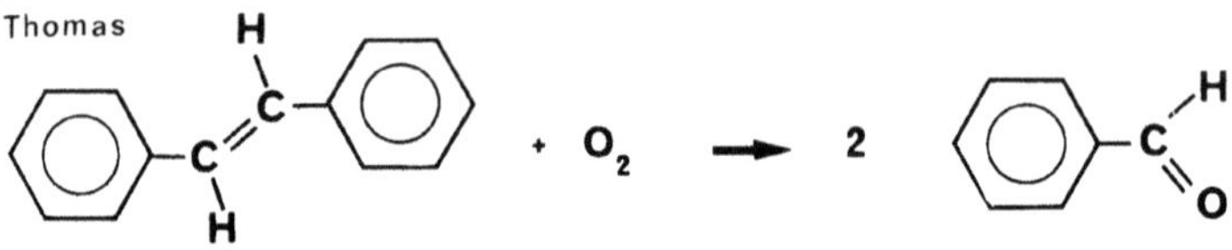


**REACTIONS OF $O_2$, NO, $SO_2$ WITH IRRADIATED SOLIDS**

Adler and al.

$$R^{\cdot} + O_2 \rightarrow RO_2^{\cdot}$$

$$R^{\cdot} + NO \rightarrow RNO^{\cdot}$$

$$R^{\cdot} + SO_2 \rightarrow RSO_2^{\cdot} \rightarrow \text{molecular products}$$

**ALKYLATION**

Lamartine and Perrin

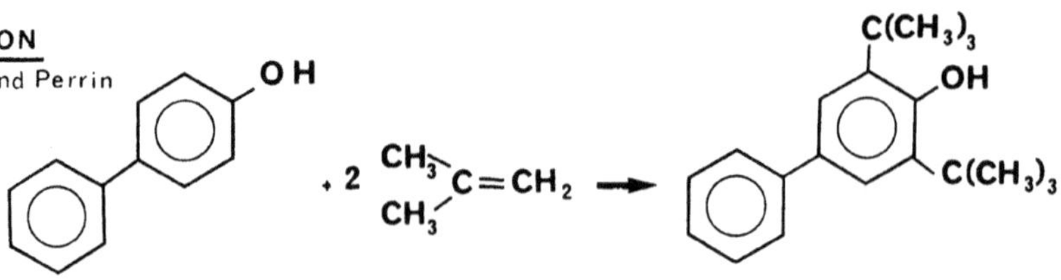


TABLE 1

**HALOGENATION**

Penzien and Schmidt

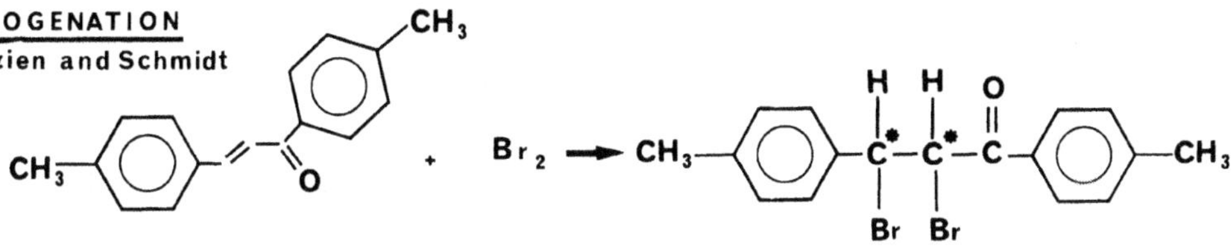


Lamartine, Perrin and al.

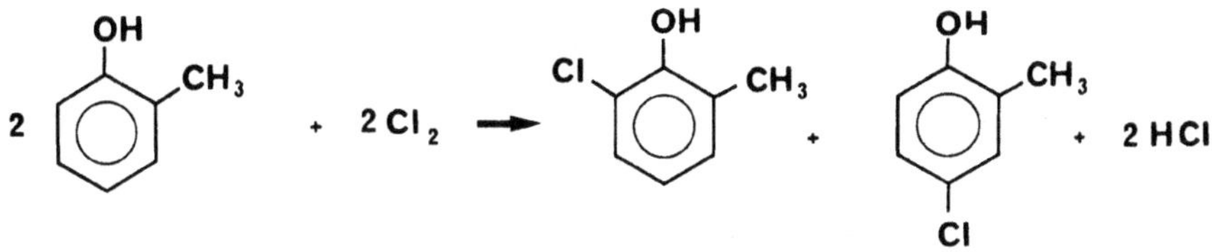


**DEHYDROHALOGENATION**

Lahav, Schmidt and al.

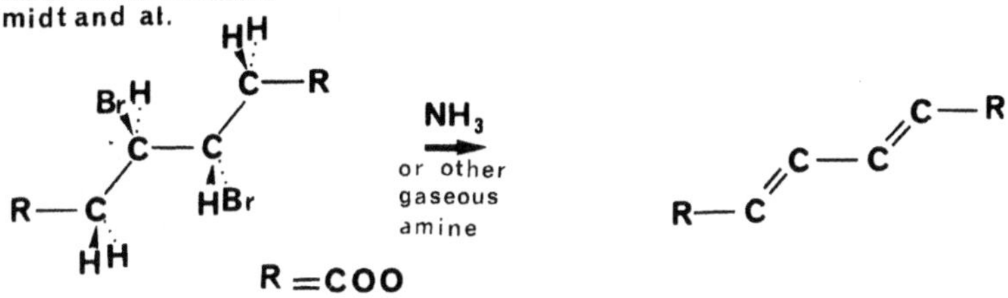


**REACTIONS WITH AMMONIA**

Miller, Curtin and Paul

**HYDROGENATION**

Lamartine and al.

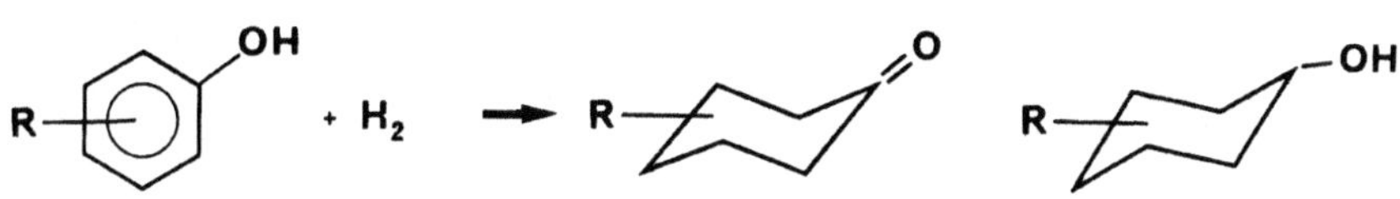


TABLE 1

a part of organic solid state chemistry which is not treated here : the photodimerisation in the crystals. From the results obtained in these reactions, G.M.J. SCHMIDT had expressed :

"There are connections between crystal structure and reactivity of crystallized organic substances and otherwise there is connection between crystal structure and the type of products obtained".

In the gas molecular organic solid reactions we must quote for example the KOLBE SCHMITT reactions (3, 4) which have been known since the last century and also still industrially used. The carboxylation of alkaline phenoxide leads to salicylaldehyde which is a promoter in the elaboration of aspirin.

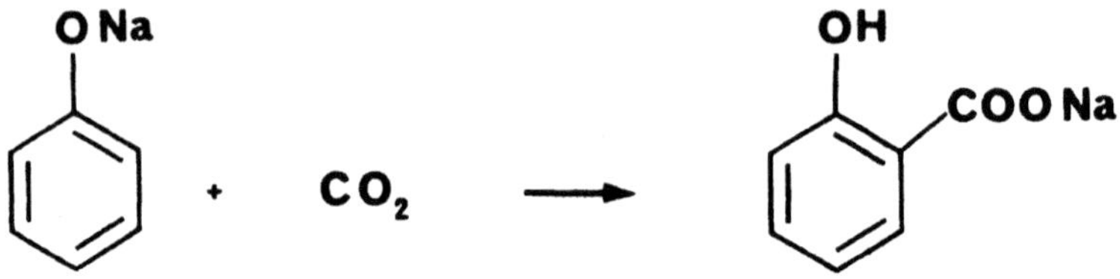


Eversince his forerunner, some workers have studied organic reactions particularly in the last years. Table I gives the main results concerning this study. We wish to report here the experimental results obtained in our group and we try to show how theoretical chemistry can assist the interpretation of such experiments.

In a first part we prove, using organic solids ground in fine powder, that the solid state is a reactive one for molecular organic compounds although this state is very often considered as invulnerable. Some catalysed or uncatalysed reactions illustrate this purpose.

In a second part, using crystals, we try to show the great interest of gas-organic crystals reactions. Taking into account the crystal structure and morphology knowledge we try to provide very powerful argument for mecanistic studies and very interesting application of these reactions.

## II. THE GREAT REACTIVITY OF ORGANIC SOLID STATE

We examine the reactivity of organic solids ground into a fine powder towards gases. We want here to emphasize the reactivity in a very mild condition of organic solids.

### II.1. The reactivity of phenoxides, significance of crystal structures

Taking up again the study of one of the first organic solid state reaction we have exhibited the surprising similarity between the action of $CO_2$ and $Cl_2$ on solid phenoxides (5, 6, 7). The figure 1 shows that in the chlorination as in the carboxylation reaction, the maximum of conversion ratio in the alcaline series is observed for the potassium and rubidium phenoxide. However we know that, in the alkaline serie, the properties change in a continuous manner from lithium to cesium.

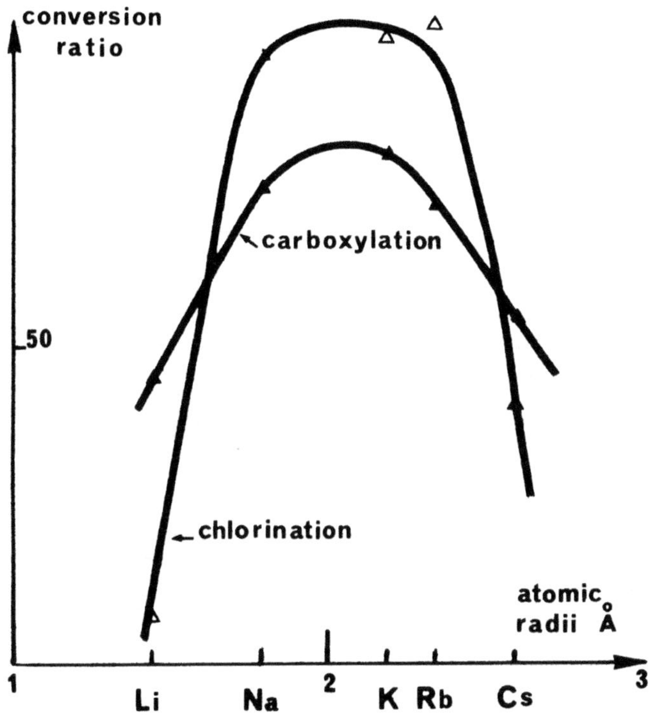


FIGURE 1
chlorination and carboxylation réactions of solid phenoxides.

Regarding the orientation of the substitution in the carboxylation presented in table II, the most likely conclusion to be closely emphasized is the dramatical decreasing of the ortho/para ratio from lithium to cesium.

| Phenoxides | Conversion ratio % | Orthohydroxy-benzoic % | Parahydroxy-benzoic % | Ortho/Para |
|---|---|---|---|---|
| Lithium | 45 | 41 | 4 | 10 |
| Sodium | 74 | 62 | 12 | 5 |
| Potassium | 80 | 48 | 32 | 1.5 |
| Rubidium | 72 | 33 | 39 | 0.85 |
| Cesium | 54 | 23 | 31 | 0.74 |

Table II. Selectivity of carboxylation of alcaline phenoxide at 150°C.

If the variation of selectivity can be explained by steric or electronic effects it is not the same concerning the conversion ratio.

It appears from this results that only the cristalline structure of phenoxides, a characteristic of the solid state, and which is still unknown for these bodies, can explain the variation observed in reactivity.

II.2. Chlorination of solids phenols, an uncatalysed solid state reaction.

Two kinds do make the extent of the studies of these reactions in our laboratory :

- this reaction can be performed without any catalyst
- this reaction takes place even at low temperature. From this fact we can make the chlorination at solid state even if the compounds melt at low temperature.

The main conclusions of these studies are as follows.

From a very full study (8) concerning 26 alkylphenols we can conclude that in the solid state at 20°C all these compounds lead, quikly, without any solvent or catalyst, to a conversion ratio from the initial phenol of 100 %. These reactions produce either chlorophenols or chlorocyclohexadienone. We have studied particularly two series of substituted alkylphenols (9). In the 3,5-dialkylphenols series the results presented in figure 2 show that the ratio of chlorocyclohexadienones increases when the substituent passes from methyl to *tert*butyl. We cannot from this experiment choose between the import of the electronic and steric effects.

Moreover in the series of 2,4,6-trialkylphenols in which all the reactive positions are blocked by an alkyl group, we observe that, for example the 2,4,6-tri*tert*butylphenol lead to the corresponding paracyclohexadienone with a conversion ratio of 92 % according to the following scheme.

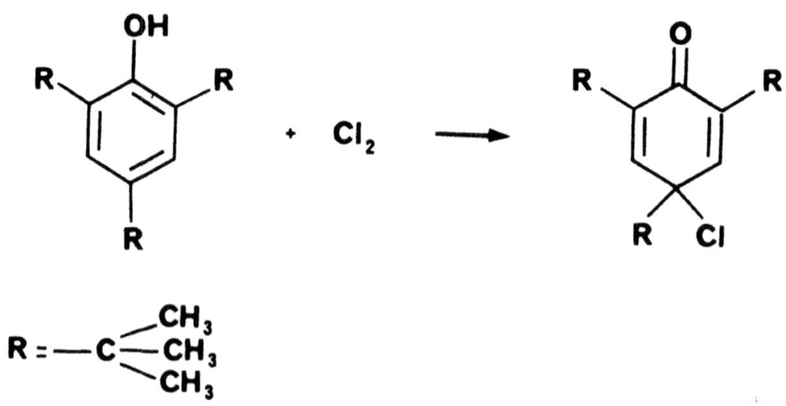

This ipsosubstitution was observed in many occasions in the solide state whereas when the reaction is performed in dissolved state this reaction does not occur.

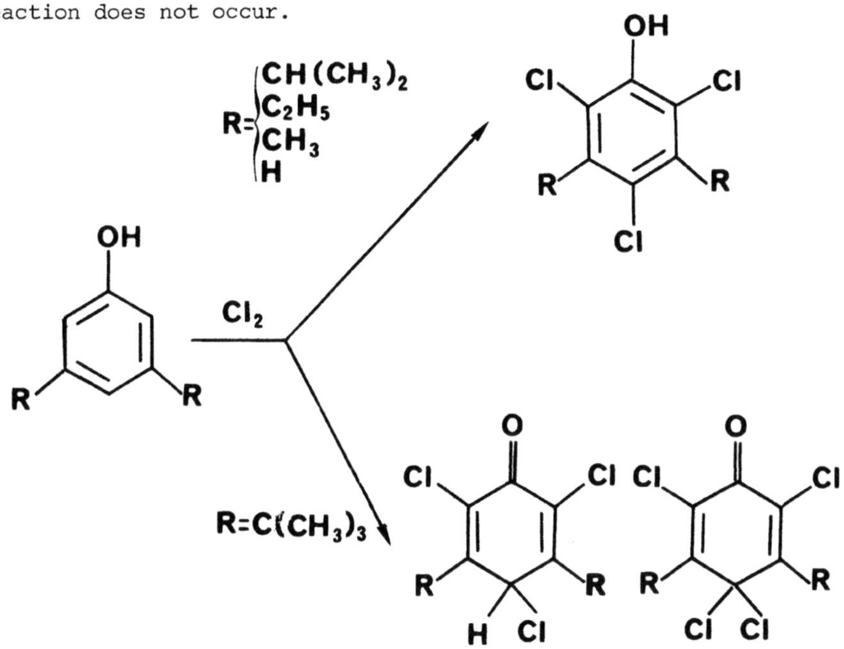


**FIGURE 2**
**chlorination of 3,5-dialkylphenols in the solid state**

In the case of 2,6-dimethyl-4-*tert*butylphenol a medium polar effect combined with the absence of steric hindrance in position 2 and 3 makes possible, as well as such ipsosubstitution in position 2 and 6, an addition reaction which leads to a chlorocyclohexenone with good conversion ratio (i.e. 95 %).

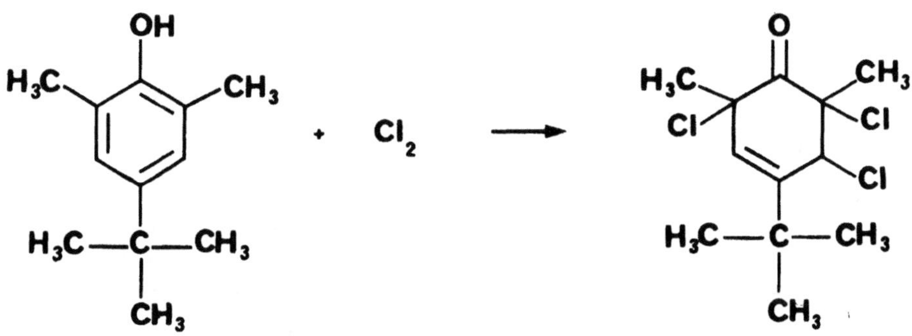

So, using the solid state, the chlorination reaction allows us to a very elegant and selective synthesis of any chlorocyclohexenone and chlorocyclohexadienone. Such compounds cannot be obtained too easily by classical procedure involving solvents.

## II.3. Hydrogenation of solids phenols and the question of the catalysed reaction in the solid state.

The chlorination reaction previously described is an uncatalysed one and we have tried to perform in the solid state a reaction using a catalyst. We have chosen the hydrogenation which is very important in the fundamental field or in the industrial point of view.

In a first serie of experiments we tried to hydrogenate the solid phenols at 1 bar of pressure and at the room temperature (10, 11, 12). Although these experimental conditions are very mild, table III shows that many phenols can react in the solid state with the hydrogen faced with catalyst.

| Compounds | Melting point (°C) | Conversion ratio |
|---|---|---|
| Phenol | 50,9 | 99 |
| 2-methylphenol | 31 | 93 |
| 4-*tert*butylphenol | 99,5 | 72 |
| 2,6-dimethylphenol | 45,6 | 94 |
| 3,5-dichlorophenol | 68 | 6 |
| 3-méthyl-6-isopropylphenol | 49,8 | 86 |
| 3-methyl-4-isopropylphenol | 111 | 20 |

Table III. Hydrogenation of solid phenols at 760 torr of initial pressure and room temperature

The catalyst used being rhodium on carbon or alumina at 5 % we can obtain a mixture of cyclohexanones and cyclohexanols.

In a second set of experiments we used various classical hydrogenation catalysts, i.e. noble metals on various support materials. The 3-methyl-6-isopropylphenol lead to the menthols or menthones stereoisomers (13). Table IV shows that all usual hydrogenation catalysts are effective in the solid state.

In the field of comparisons of solid state and dissolved state we can observe that the alcohols as solvents cannot allow to the hydrogenation reaction although the aprotic polar and non polar solvent such as cyclohexane or 1,2 dichloro-ethane can be used satisfactorily.

Concerning the stereoselectivity of this solid state reaction we can notice (10, 12) that the catalyst support plays an essential role. The hydrogenation of 4-tertbutylphenol (figure 3) with Rh/C at 5 % gives a mixture of 4-tertbutyl cyclohexanols with a high ratio of cis isomers

| Catalyst | Convertion ratio (%) | Reaction products | | | | | |
|---|---|---|---|---|---|---|---|
| | | 1 | 2 | 3 | 4 | 5 | 6 |
| Pd/C | 50 | 14 | 86 | 0 | 0 | 0 | 0 |
| $Pd/Al_2O_3$ | 26 | 17 | 83 | 0 | 0 | 0 | 0 |
| Rh/C | 77 | 10 | 38 | 9 | 42 | 0 | 1 |
| $Rh/Al_2O_3$ | 81 | 2 | 17 | 9 | 54 | 11 | 7 |
| Pt/C | 99 | 1 | 8 | 12 | 42 | 29 | 8 |
| $Pt/Al_2O_3$ | 87 | 1 | 5 | 11 | 66 | 12 | 5 |
| Ru/C | 20 | 2 | 16 | 3 | 61 | 0 | 18 |
| $Ru/Al_2O_3$ | 0 | 0 | 0 | 0 | 0 | 0 | 0 |
| $PtO_2$ Adams | 21 | 25 | 75 | 0 | 0 | 0 | 0 |
| Ni Raney | 20 | 0 | 1 | 6 | 75 | 9 | 9 |

1 : menthone 2 : isomenthone 3 : néomenthol
4 : néoisomenthol 5 : menthol 6 : isomenthol

Table IV. Solid state hydrogenation of thymol (3-methyl-6-isopropyl-phenol).

while $Rh/Al_2O_3$ at 5 % gives in the same conditions a mixture containing approximatively the same quantites of cis and trans isomers.

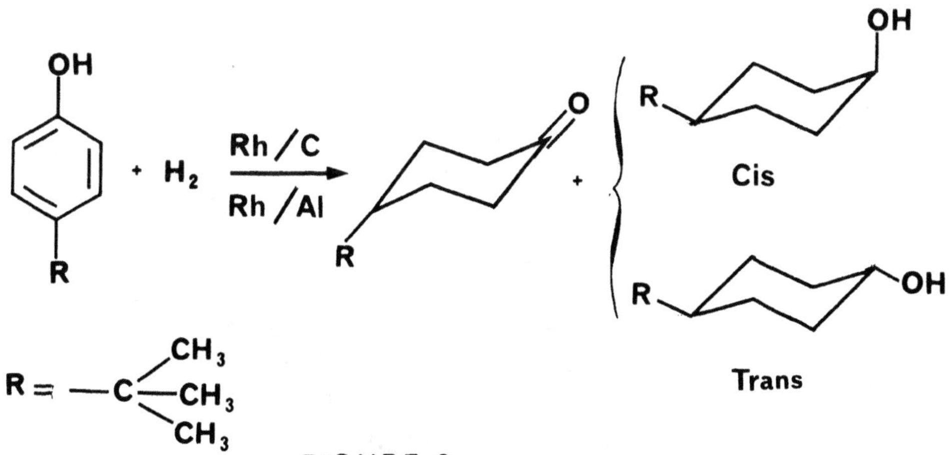


FIGURE 3

**Solid state hydrogenation of 4-tertbutylphenol**

In all these hydrogenation reactions we must notice that the experimental condition cannot perform very good contact between catalyst and reactants. Although the phenol ground in a fine powder is simply mixed with catalyst only with the aid of a spatula, the hydrogenation takes place. It seems that the high conversion ratio observed can be explained by assuming the presence of "activated hydrogen" at some distances from the catalyst. This hypothesis is directly related to the hydrogen "spillover" effect evidenced in heterogeneous catalysis.

## III. THE SIGNIFICANCE OF GAS-ORGANIC CRYSTALS REACTIONS

After the study showing the reactivity of the organic solid state, we examine how it is interesting to perform gas reaction with organic single crystals. We hope to relate the obtained results particularly reactivity and selectivity with the crystal structures. From this point of view we chose early crystals and reactant to observe major difference either in reactivity of in selectivity after we would investigate more and more fine differencies.

### III.1. Two crystals of the same compounds : polymorphic variety.

Many parasubstituted phenols exhibit two or more polymorph forms (14) whose crystal structures are known. Is their reactivity identical towards a same gaseous reagents ?

The reaction (figure 4) between chlorine and 4-chlorophenol (15) allows us to evidence the two following results :

- metastable crystals react more quickly as those of the stable form
- microcristalline powder obtained from the two forms cannot exhibit differencies.

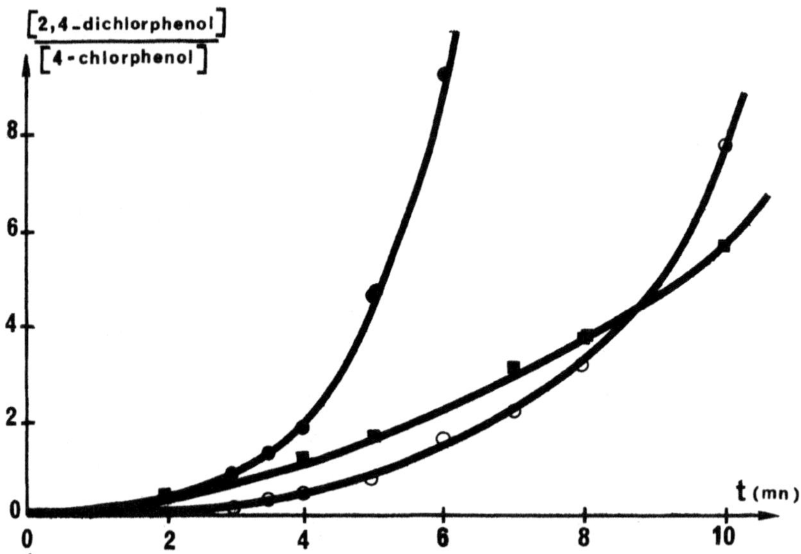


(● métastable crystals, O stable crystals, ■ stable and metastable powder)

FIGURE 4

**Reaction between chlorine and 4-chlorphenol**

Crystal structure examination related to morphology shows that the orientation of molecules are not identical in these two forms.

In the metastable form all the molecules can react by an ortho position pointed towards the external faces of the crystal. On the contrary in the stable form crystals, many molecules exhibit towards the external faces either chlorine atoms or meta position which are both unreactive. As the crystals are ground in fine powder appears in statistical way all the possible directions. So we cannot observe differencies in the reactivity for the two powders proceeding from the two crystalline forms.

These results are in very good agreement with the view point of M.D. COHEN, G.M.J. SCHMIDT (16) and H. MORAWETZ (17) : If compounds exist in various polymorphic forms, they do not react at the same rate.

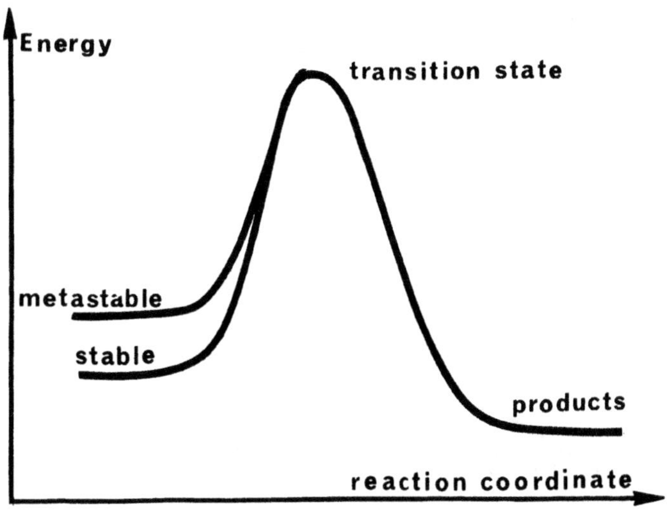


**FIGURE 5**
**Diagram for a reaction of two polymorphic crystals**

From these experiments we can ask in the theoretical point of view (figure 5) how a stable form is reactive with respect to a metastable one ? Since the reaction products are identical and the energy differences between the two forms are known, is the transition state the same ? In this case the reactivity differences are only due to the energy difference between the two forms.

III.2. The same crystal but two type of faces.

We can link together chemical reactivity, crystal structure and morphology using stable single crystals of 4-chlorophenol (18). In this crystal (figure 6) we have two sets of faces in zone (i.e. parallel to $\vec{a}$ axis or $\vec{c}$ axis) that we can put in reaction successively (19).

The following table V shows that the faces parallel to c axis react more slowly than the faces parallel to $\vec{a}$ axis.

| Crystal | Masse of crystal (mg) | Reacting faces | Conversion ratio 1 | Conversion ratio 2 |
|---|---|---|---|---|
| $A_1$ | 1984.2 | $//\vec{c}$ | 15.7 | 15.4 |
| $A'_1$ | 2053.3 | $//\vec{a}$ | 23.3 | 22.8 |
| $A_2$ | 8943.6 | $//\vec{c}$ | 9.8 | 9.8 |
| $A'_2$ | 9107.5 | $//\vec{a}$ | 14.7 | 14.5 |

Table V. Reactivity of two sets of faces for 4-chlorophenols crystal
1) Conversion ratio obtained by weighing of crystal before and after reaction.
2) Conversion ratio obtained by chromatographic determination.

In this case too the reactivity differencies can be explained by considering the crystal structure (20).

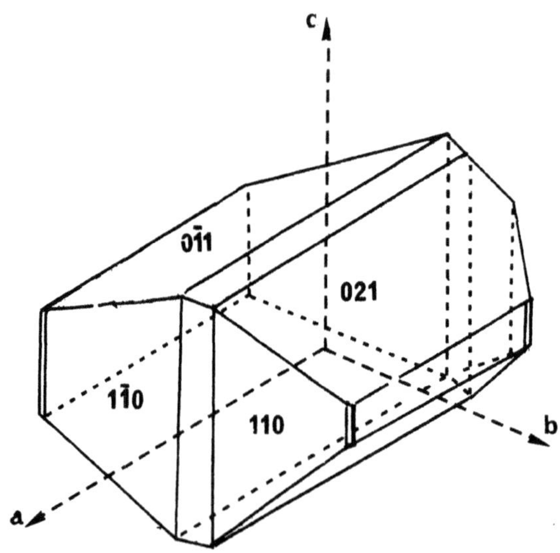


**FIGURE 6**
**Single crystal of 4-chlorophenol (stable form)**

The set of faces as (021) presents molecules for which an ortho position is accessibles to gaseous reactants while the set of faces as (100) presents molecules with chlorine atom or carbon in meta position which are not reactive. Thus the faces parallel to $\vec{c}$ axis react more slowly than those parallel to $\vec{a}$ axis.

III.3. Quantitative study of selectivity using the same crystal but two perpendicular direction.

For these studies (21) we chose the 2-methylphenol which can react in 4 position (para) or in 6 position (ortho) towards an electrophile agent as chlorine. A typical crystal of 2-methylphenol is a rod extended along the $\vec{c}$ axis. In this crystal we cut plates of 1 or 2 mm thickness some parallel to $\vec{c}$ axis others perpendicular to the $\vec{c}$ direction. These plates are submitted to the action of chlorine and the reaction products are analysed by gas chromatography. The ratio 6-chloro-2-methylphenol/4-chloro-2-methylphenol, so called o/p ratio, provide the selectivity of the reaction. Whatever the conversion ratio the selectivity is always greater for the plates cut perpendicularly to $\vec{c}$ axis than these obtained for the plates cut parallel (figure 7).

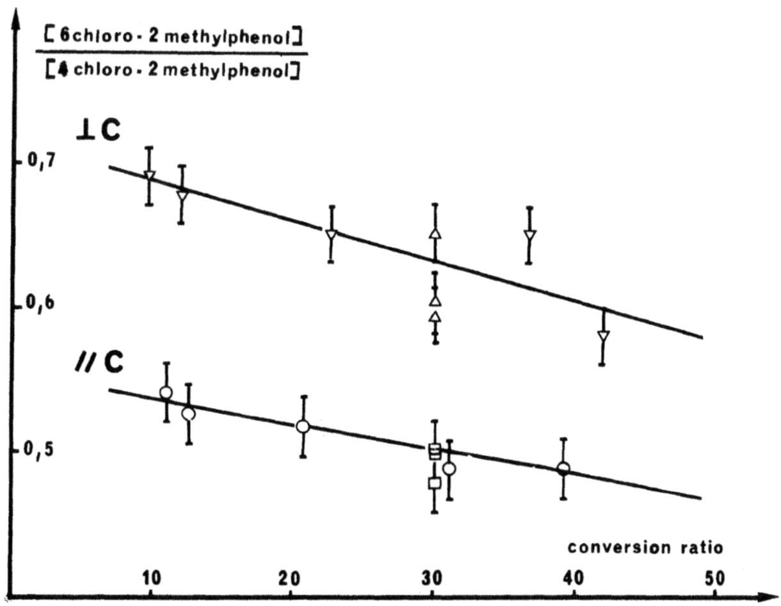


FIGURE 7
Plot of o/p ratio versus conversion ratio
(error bars indicate analytical errors).
Reprinted with permission from J.A.C.S. 99. 5436, 1977. Copyright 1977 American Chemical Society.

These surprising experimental results cannot be explained by the irregularities on the faces because of the fact that this variation of selectivity is independent from the conversion ratio. These results can be understood if we consider the orientation of 2-methylphenol molecules within the crystal faces and the accessibility of reaction sites. On the plates cut perpendicularly to the $\vec{c}$ axis the molecules are exposed as in the figure 8 and the accessibility to the ortho or paraposition is the same. On the contrary on the plates cut parallel to the $\vec{c}$ axis 1/3 of the molecules present either the paraposition or an unreactive position such as OH or $CH_3$ group. On the stereoscopic wiew of figure 8 the marked molecules (*) are in these situation towards the chlorine approach on the faces (11$\bar{2}$0). This argument is also valid for the attack on the faces (10$\bar{1}$0) or (01$\bar{1}$0). This peculiar orientation for 1/3 of molecules could favour the accessibility of paraposition and lead to a lower o/p ratio for the plates cut parallel to the $\vec{c}$ axis than for those cut perpendicular to the $\vec{c}$ axis.

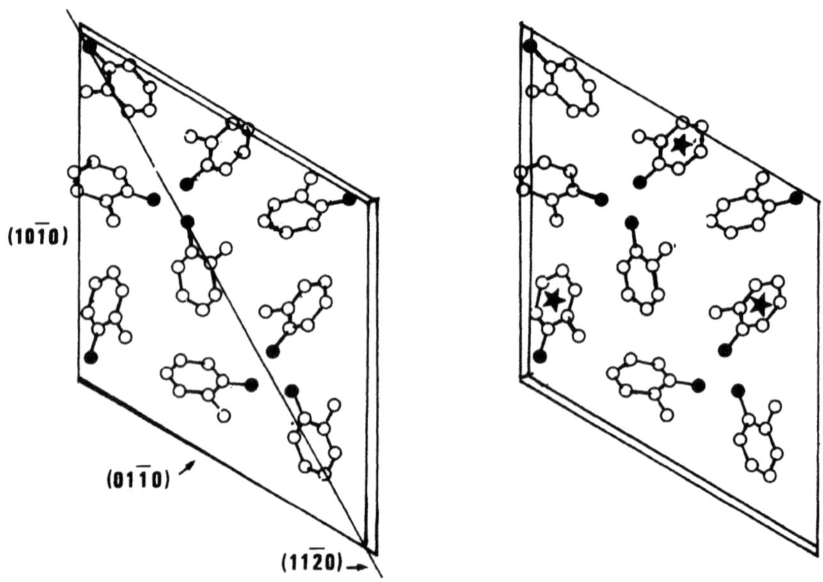


FIGURE 8

**Stereoscopic view of the 2-methylphenol crystal structure (● = oxygen , ○ = carbon)**

Reprinted with permission from J.A.C.S.99,5436,1977.Copyright 1977 American Chemical Society.

Are these results meaningful ? We think that three requirements are necessary.

- The diffusion must not be significant in the reaction
- The thermal effect must remain low

- The reaction proceeds trough two independent reaction paths one corresponding to the orthosubstitution and the other to parasubstitution.

- Diffusion

If the gaseous molecules of chlorine can diffuse inside the bulk of crystal, the selectivity cannot be ruled by the presentation of molecules on the crystal faces. For the reaction between the chlorine and the 2-methylphenol the diffusion of chlorine is highly improbable since the minimum dimension of a chlorine molecules evaluated by Van der Waals contact radii is about 3.6 Å and the largest available channel on the structure are about 3.1 Å.

In the case of reaction with hydrogen the very low size of molecules cannot allow us to show reactivity or selectivity differences on various crystal faces. According to these reasons we have already attempted such hydrogenation experiment without success.

- Thermal effect

The thermal effect of reaction must remain low otherwise else the evolved heat leads us to a local disorder which gives non-significant results.

- Two independent reaction paths

If one transition state ($\pi$ complexe) leads at the same time of the ortho and parasubstitution products, the o/p ratio is fixed and cannot be changed by various experimental conditions particularly the exposed crystal face.

The anisotropy observed for this gas-organic solid reaction suggests that the chlorination of 2-methylphenol proceeds through two reaction paths each having a transition state, one corresponding to orthosubstitution, and the other to parasubstitution. Then, it is worth checking if these conclusions are consistent with theoretical chemistry calculations.

A straightforwards pattern of the chlorine approach direction is put on figure 9b and leads us to the energy calculations using the CNDO/2 (22) method. The curves of figure 9a show the existence of a stable complex towards a vertical approach on the center of aromatic nucleus at d ≅ 4 Å. This complex ($\pi$ complex) is similar to those studied by HASSEL and STROMME (23) between halogens and benzene molecules. Both approaches 1 and 3 exhibit two different potential barriers and are more favoured than approach 2.

This very simplistic model must be now studied with more sophisticated method using energy minimisation process versus geometry of complex. Afterwards it needs taking into account the environments of the reacting molecules to study his reactivity.

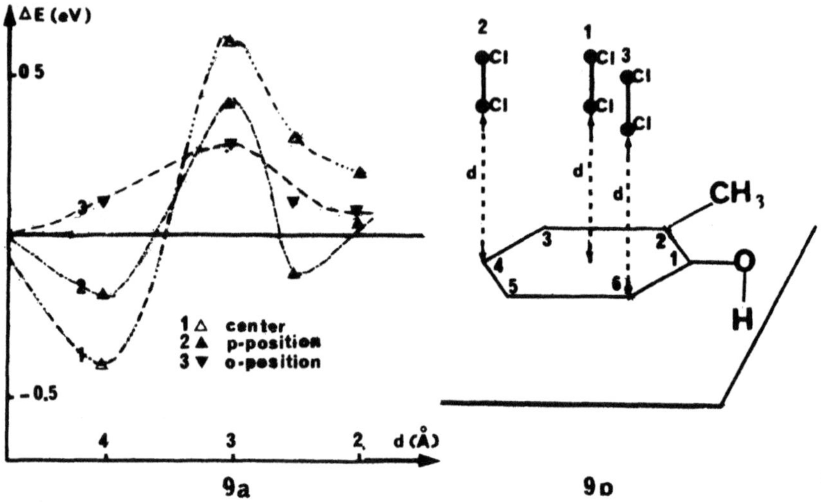


**FIGURE 9**
**Theoretical study of chlorine-2-methylphenol reaction.**
ΔE = E(supersystem) − E(isolated molecules)
Reprinted with permission from J.A.C.S. 99, 5436, 1977. Copyright 1977 American Chemical Society.

In this way we have studied two points view to determine the signifi cance of lattice constraints both on the conformations and on the molecular geometry deformation. On the series of 3-methylisopropyl-phenols isomers (24) we compare the conformation in the solid state determined by X ray diffraction with those theoretically calculated for isolated molecules in their ideal gaseous state. We showed in this paper the very good agreement between the calculated and observed conformers and we can conclude to the very low importance of the lattice containts in our crystals.

On the other hand, the survey of the crystal structure of phenolic compounds (about 50 compounds) shows systematic deformation of the aromatic nucleus and around the hydroxyl group according to the figure 10.

Molecular orbital calculation (25) with geometry optimisation by MNDO method shows that despite the fact they are supposed to be related to the isolated molecules, the results are in good agreement with crystal structure determination. Thus, in our crystals the conformations and the molecular geometry deformation are intrinsic properties of molecules, unmodified by the lattice constraints. We must now set out the calculation of the reactivity of the molecules perturbed by these neighbourhoods. In this way the method proposed by CLAVERIE (26) for molecular crystals will allow us to carry out this work.

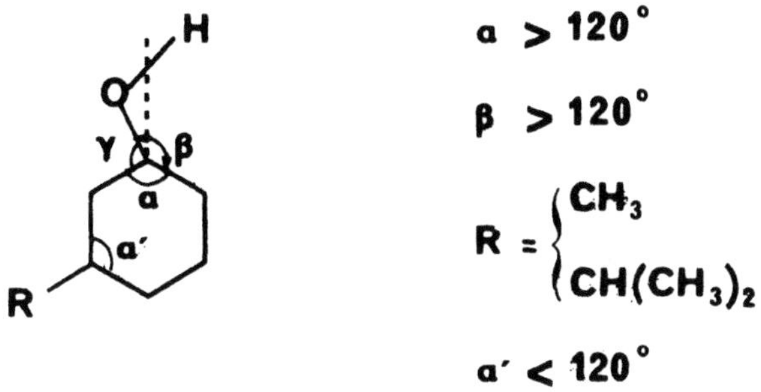


**FIGURE 10**
**Deformation of the aromatic nucleus in phenolic compound**

III.4. The same crystal, the same direction but top and botton reaction.

Some crystals crystallize in polar groups. These crystals exhibit a direction whose end is not symetrically related to its oppositely directed end. For these reactions we have selected (27) the 3-methyl-4-isopropylphenol which crystallises in the enantiomorphous groups $P4_1$ or $P4_3$. In this noncentrosymetric group the molecules form by hydrogen bond, infinite chains around the polar 4-fold axis, we can find left crystals and right crystals and we know how to growth. These crystals can be recognized with polarising microscope assistance.

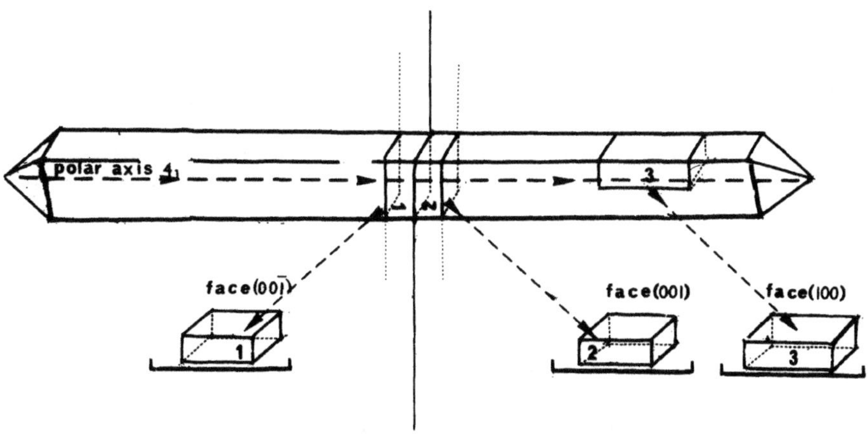


**FIGURE 11**
**Diagram showing plates orientation for reaction of polar crystal.**

If, as in figure 11, in a 3-methyl-4-isopropylphenol crystal we cut plates perpendicular to the 4-fold axis, the two oppositily directed (001) and ($00\bar{1}$) faces are cristallographycally different. We purpose to obtain chemical evidence of this phenomenon.

The two plates, cut as in figure 11 do not turn, are submitted to the chlorine action in a thermoregulated reactor with a plate cut parallel to the axis as control. After reaction the three plates were removed and weighed. The results are collected in the following table.

| Experiments | (001) face perpendicular up | ($00\bar{1}$) face perpendicular down | (100) face parallel |
|---|---|---|---|
| 1 | 2.5 | 1.3 | 1.3 |
| 2 | 1.3 | 0.7 | 0.8 |
| 3 | 2.1 | 1.4 | |

Table VI. Weight increase (mg) observed after chlorination reaction of various plates of 3-methyl-4-isopropylphenol.

These experiments emphasize that one perpendicular face react more quickly than the other (approximatively two time) and more quickly than the plates cut parallel to the 4-fold axis. In a word these simple studies exhibit reactivity discrimination which is due to the polarity of the 4-fold axis. We are actually performing such experiment on other polar crystals to generalize this result.

III.5. Use of the chirality of achiral molecules arrangement : asymetric synthesis possibility.

We take advantage of the possibility to carry hydrogenation reaction in the solid state and of the existence of chiral crystals to perform an asymetric synthesis. The achiral compounds 3-methyl-4-isopropylphenol crystallize in the enantiomorphous space group $P4_1$ or $P4_3$. A left or right single crystal is ground in fine powder, mixed with a classical hydrogenation catalyst like Pt/C and submitted to hydrogen to obtain alkylcyclohexanols and alkylcyclohexanons mixtures.

After reaction the products are removed from the catalyst by centrifugation. With left crystal powder we obtain in several experiment a low rotatory power. Note that this rotatory power is time stable.

This new result must be confirmed by identification in the reactionnel mixture of the compounds in enantiomeric excess. This reaction of achiral compounds with achiral catalyst leading to rotatory power is an assymetric synthesis using only the chirality of crystalline arrangment.

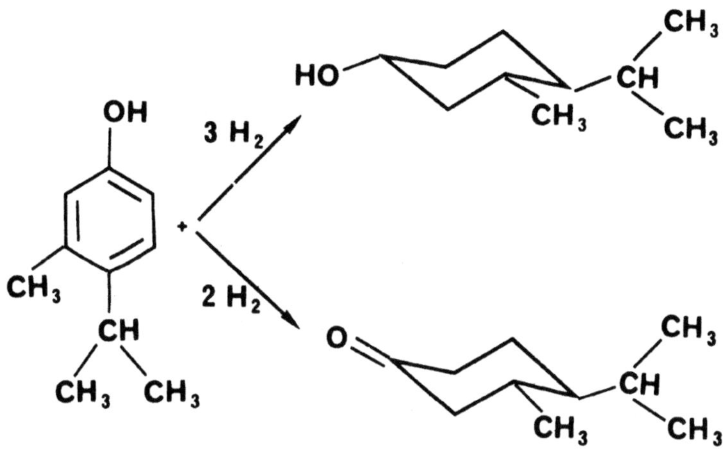


IV. CONCLUSIONS

Performing reactions in the organic solid state which is very reactive can be interesting on several accounts.

In the practical organic chemistry field, numerous advantages result from the use of such reactions.

- to perform organic reactions without any solvents or catalysts can be a most striking interest even industrial purpose.

- organic solid state synthesis allows us to new reactions. For example ipsosubstitution at carbon bearing an alkyl radical or selective production of cyclohexenone during solid phenol chlorination.

An other staighforward reason to carry out oragnic reaction in the solid state is the possibilities to obtain reaction in the very mild energetic conditions. We can for example phenols hydrogenation in the solid state at room temperature and at one bar of initial pressure. It appears then, that we have only just begun to exploit these reactions to obtain the thermodynamically instable isomers.

Moreower the organic solid state chemistry allow us the realisation of an absolute assymetric synthesis from achiral reactants and catalyst using only the chirality of crystalline arrangement.

In the field of theoretical chemistry and in the reaction mechanism studies we have emphasized the importance of the structural presentation of molecules on the crystal faces. This accessibility for the gaseous reactant towards the molecules exposed on the crystal faces seems to govern both the selectivity and the reactivity of organic solid state. Indeed, if in the melts, in solution or in the gaseous state the reactant be freely oriented, it cannot be the same situation in the gas-solid reactions. In these reactions one of the reactants is in a particular organised state, well know by crystallographic studies. In the recent years the availability of crystal structures for molecular crystal have know a very rapid development. It appears that this organised state is interesting for mechanism studies.

The usual method for studying the reactivity with theoretical calculation involve the isolated molecules and in the recent years the molecule perturbated by solvent. For the gas-solid reaction we want to take into account the surrondings perturbation of molecule in the crystal or at the crystal surface. The first studies have shown the contribution of theoretical chemistry to the knowledge of these perturbations and their effects on the molecular deformation or an the molecular conformation. With more sophisticated method we can seen how the crystal may play a part in determing the reactivity of molecules.

This paper report the experimental results obtained in a interdisciplinary group (ERA 600 of CNRS) including chemists crystallographers and theoretical chemists.

## REFERENCES


1. Cohen, M.D., Schmidt, G.M.J., 1964, J. Chem. Soc., pp. 1996-2000.
2. Morawetz, H., 1963, Physics and Chemistry of the organic solid state (Ed. D. Fox, M.M. Labes, A. Weissberger), Interscience Publ., New York.
3. Kolbe, H., 1874, J. Pr. Chem., (2)10, pp.89-112 and 1875, (2)11, pp. 24-26.
4. Schmitt, R., 1885, J. Pr. Chem., (2)31, pp. 397-411.
5. Berny, M.-F. and Perrin, R., 1967, C.R. Acad. Sci., 265, pp. 492-493.
6. Lamartine, R., Vincent-Falquet-Berny, M.-F. and Perrin, R., 1972, Third Int. Symp. "Chemistry of the organic solid state", Glasgow.
7. Berny, M.-F. and Lamartine, R., 1975, Bull. Soc. Chim., 1-2, pp. 47-52.
8. Lamartine, R. and Perrin R., 1974, J. Org. Chem., pp. 1744-1748.
9. Bertholon, G., Lamartine, R., Perrin, R., Vincent-Falquet, M.-F., 1971, Groupe d'Etude des Méthodes Thermochimiques, Nice.
10. Repellin, M., Lamartine, R., Perrin, R. and Bertholon, G, 1977, C.R. Acad. Sci., 285, pp. 109-111.
11. Repellin, M., Perrin, R., Lamartine, R., Bertholon, G. and Pajonk, G., 1977, C.R. Acad. Sci., 285, pp. 335-336.
12. Lamartine, R., Bertholon, G., Perrin, R., Perrin, M., Thozet, A., 1979, Mol. Cryst. Liq. Cryst., 52, pp. 293-302.
13. Lamartine, R., Perrin, R. and Bertholon, G., 1980, C.R. Acad. Sci., 291, pp. 219-221.
14. Perrin, M., Michel, P. and Perrin, R., 1975, J. Chim. Phys., 7-8, pp. 851-854.

15. Perrin, M., Michel, P., Lamartine, R., Perrin, R., Sept. 1973, First European crystallography Meeting, Bordeaux.
16. Cohen, M.D. and Schmidt, G.M.J., 1961, Reactivity of Solids, J.H. de Boer, Elsevier, pp. 556-562.
17. Morawetz, H., 1966, Science, 152, pp. 705-711.
18. Michel, P., Perrin, M., Perrin, R., Bavoux, C., 1972, Crystallogenese Experimentale, Rennes.
19. Lamartine, R., Perrin, R., 1974, C.R. Acad. Sci., 279, pp. 367-369.
20. Perrin, M., Michel, P., 1973, Acta Cryst., B29, pp. 253-258.
21. Lamartine, R., Perrin, R., Bertholon, G., Vincent-Falquet, M.-F., 1977, J. Amer. Chem. Soc., 99, pp. 5436-5438.
22. Pople, J.A., Santry, D.P., Segal, G.A., 1965, J. Chem. Phys., 43, pp. 5129-5135.
Pople, J.A., Segal., G.A., 1965, J. Chem. Phys., 43, pp. 5136-5141 and 1966, J. Chem. Phys., 44, pp. 3289-3296.
23. Hassel, O., Stromme, K.O., 1958, Acta Chem. Scand., 5, pp. 1146-1147 and 1959, Acta Chem. Scand., 9, pp. 1781-1786.
24. Bertholon, G., Thozet, A., Perrin, R., Decoret, C., 1981, J. Mol. Struct. Thermochem., 76, pp. 313-321.
25. Perrin, M., Thozet, A., Bertholon, G., Decoret, C., Royer, J., 1981, J. Mol. Struct., 70, pp. 87-94.
26. Bertholon, G., Perrin, R., Lamartine, R., Thozet, A., Perrin, M., Caillet, J., Claverie, P., 1979, Mol. Cryst. Liq. Cryst., 52, pp. 285-292.
27. Thozet, A., 1981, Thesis, Lyon.

# SOME THEORETICAL QUESTIONS CONCERNING THE MECHANISM OF FISCHER-TROPSCH SYNTHESIS.

I.G. Csizmadia
Department of Chemistry, University of Toronto,
Toronto, Ontario, Canada M5S 1A1

Catalytic hydrogenation of carbon monoxide, irrespectively whether it leads to the formation of methanol, methane or higher hydrocarbons, is one of the most important reactions of our times. The methanol

$$CO + H_2 \xrightarrow{\text{methanol synthesis}} CH_3OH \quad [1]$$

$$CO + H_2 \xrightarrow[\text{Sabatier-Senderens synthesis}]{\text{methanation}} CH_4 + H_2O \quad [2]$$

$$CO + H_2 \xrightarrow{\text{Fischer-Tropsch synthesis}} C_nH_{2n+2} + nH_2O \quad [3]$$

synthesis [1] is a widely used industrial process yet methanation (the Sabatier-Senderens reaction[1]) [2] and the production of synthetic gasoline (the Fischer-Tropsch synthesis[2]) [3] just becoming to be considered important again. Yet the mechanisms of these reactions are far from being understood.

The balanced equation for the Sabatier-Senderens reaction is

$$CO + 3H_2 \rightarrow CH_4 + H_2O$$

leading to an exothermic reaction of the following magnitude

$$\Delta H \simeq -50 \text{ kcal/mole}$$

The corresponding relations for the Fischer-Tropsch synthesis are the following :

$$nCO + (2n+1)H_2 \rightarrow C_nH_{2n+2} + nH_2O$$

$$\Delta H \simeq -(11 + 39n) \text{ kcal/mole}$$

Although the thermochemistry of these reactions are reasonably clear, their mechanism, that is also related to the energies of activation, are far from being understood.



*R. Daudel, A. Pullman, L. Salem, and A. Viellard (eds.), Quantum Theory of Chemical Reactions, Volume III, 77–84.*

A number of mechanistic views have been advocated during the past half a century or so but many of them proved to be unacceptable.

The recent work of Pettit and coworkers[3] shed some light on the latter part of the mechanism but early part of the mechanism is still that proposed by Fischer and Tropsch :

[4]

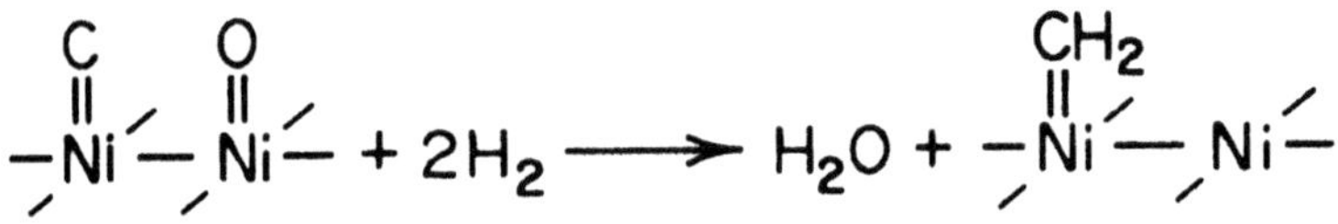

[5]

According to Pettit the mechanism includes chain initiation, chain propagation and chain termination :

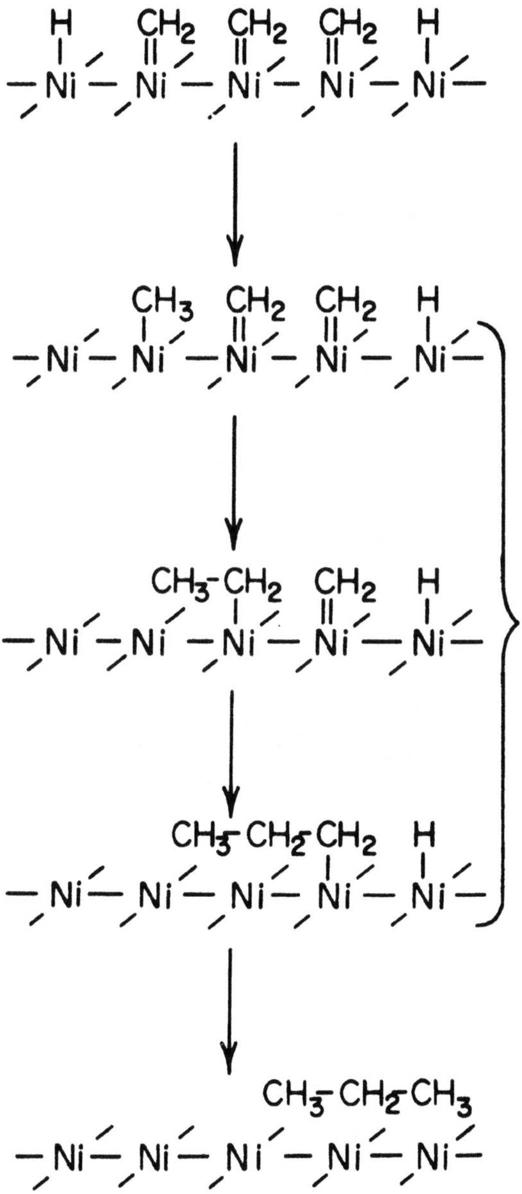

Chain Initiation [6]

Chain Propagation [7]

Chain Termination [8]

When nickel is used as a catalyst the major product is $CH_4$ (i.e. the number of chain propagation steps is zero). Using Ni-Fe as catalyst higher hydrocarbon homologs are obtained ; thus the number of chain propagation steps is greater than zero. It Cu is present than the chain termination step is different.

[9]

leading to the formation of a carbonyl compound-Cu-complex that may be further hydrogenated to the corresponding aldehyde or alcohol.

Although we now have some understanding of the later portion of mechanism, the early part of the mechanism still has a number of unanswered questions.

First of all it is not clear if the first reaction, i.e. [4], is exothermic at all. In order to answer this question one would need at least experimental bond dissociation energies. The bond dissociation energy of NiO has already been reported[4] to be of 86.5±5 kcal/mole however, no corresponding value was reported for NiC. As even the existence of NiC has not been reported as yet in the literature so one may only wonder about its stability. However, a number of heteronuclear Ni diatoms have been investigated involving elements of the upper right hand corner of the periodic systems. The bond dissociation energies ($D_0$, in eV) for those diatoms that Herzberg reported[5] are truncated arbitrarily at one decimal figure and are given in the following scheme

| | | | |
|---|---|---|---|
| C | N | O<br>3.8 | F |
| Si<br>3.2 | P | S<br>3.5 | Cl<br>3.8 |
| Ge<br>2.8 | As | Se | Br<br>3.6 |
| | | | I<br>3.0 |

[10]

Clearly one can only make a rough extrapolation for C but it seems reasonable to anticipate its value to be 3.5 eV corresponding roughly to an 80 kcal/mole NiC bond dissociation energy.

In order to see some of the details of the reaction energetics one needs incorporate the atomization of $H_2$ on Ni. The bond dissociation energy of Ni-H is reported[5] to be 71 kcal/mole thus the process :

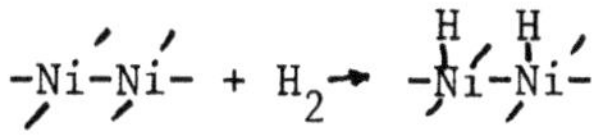

[11]

may be estimated to be exothermic to the following degree :

$$\Delta H \simeq -2D_o(Ni\text{-}H) + D_o(H_2) = -38 \text{ kcal/mole}$$

Thus the hydrogenation reaction of CO assuming that C≡O dissociates on the Ni surface to form NiC and NiO can be depicted as shown by Figure 1. Clearly the dissociation of $H_2$ on Ni is exothermic but the dissociation of CO is endothermic. However the 1:3 mixture of CO and $H_2$ necessary for the methanation is still exothermic by about -24.5 kcal/mole.

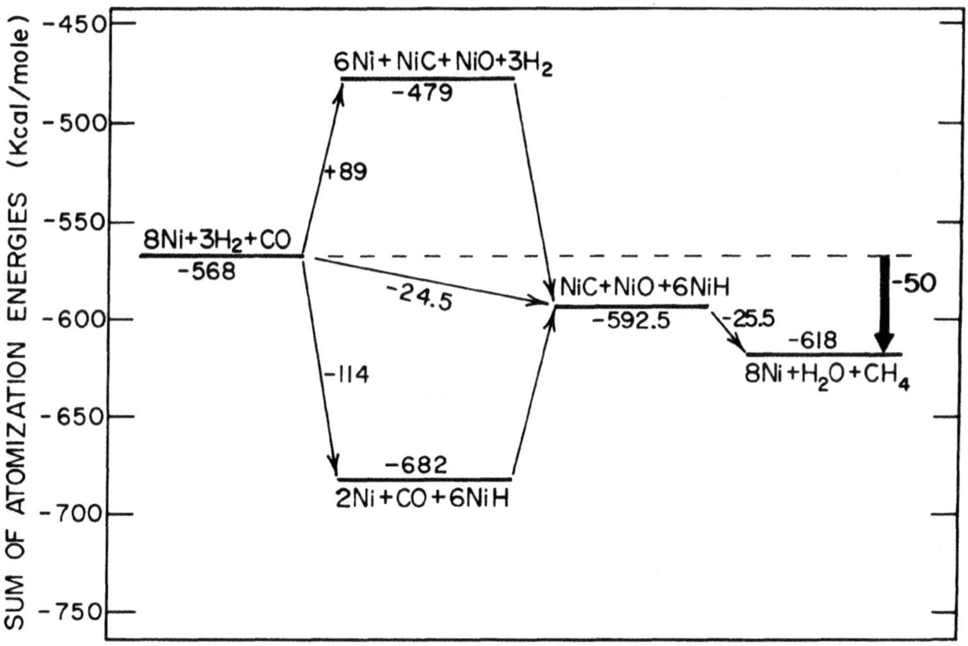


Figure 1. Thermochemistry of Methanation

The thermochemistry of the intermediates shown in Figure 1 is based on the assumption that the binding energy of C and O to the Ni surface is the same as that in NiC and NiO respectively. If other bonding arrangements occur, such as those shown below

[12]

then of course even the dissociation of CO on the Ni surface may be slightly exothermic.

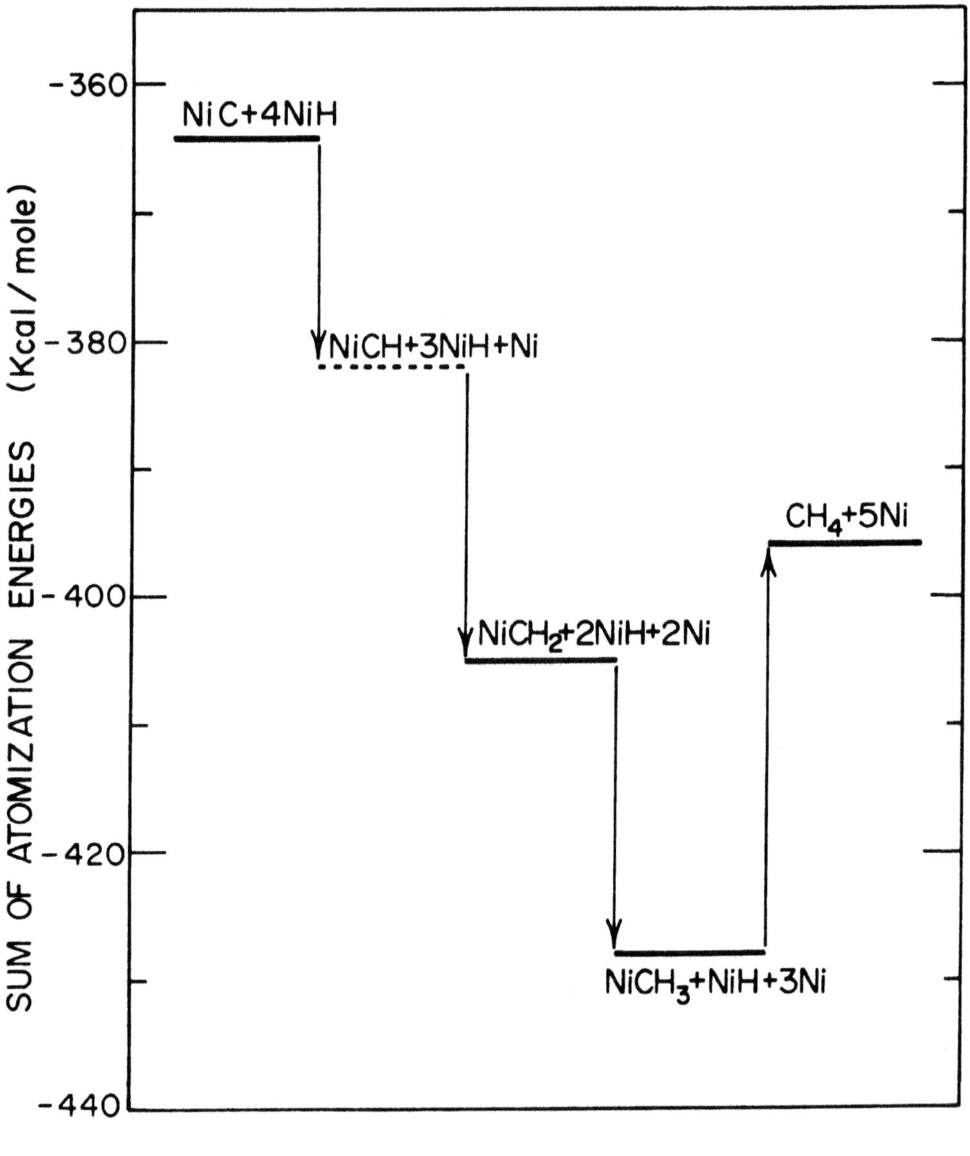


Figure 2. Thermodynamics energy levels associated with the stepwise hydrogenation of NiC.

At any rate no mechanistic information can be derived from these equations concerning the formation of the key intermediate $NiCH_2$. The idea of hydrogenating NiC that is formed initially is thermodynamically acceptable as illustrated in Figure 2, but no proof is available at this stage if this is exactly the way the reaction occurs.

For the early part of the reaction two rival mechanisms were suggested. One was the "carbide mechanism" as discussed above and the other one was the "hydroxycarbene mechanism" that was thought to involve the following intermediate :

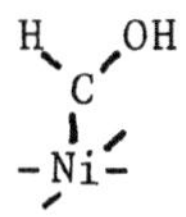

However, there is no need to be convinced that definitely one or the other mechanism is operative. In principle one may devide the individual reactions into two types : dissociation and hydrogenation ; the available combinations of these two types are summarized in Figure 3.

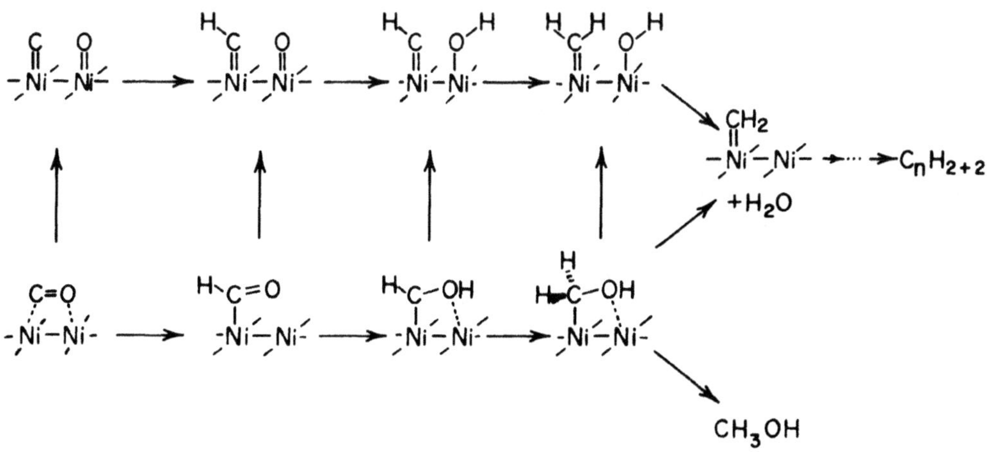


Figure 3. Reaction mechanism network for the hydrogenation of CO. (Vertical arrows indicate dissociative mode, horizontal arrows indicate hydrogenation mode. The last steps marked as →...→ are the reactions given by equations [6], [7] and [8]).

One can view this figure as an approximate topological pattern of a reaction hypersurface the structures represent minima while the arrows represent paths between minima. Then two rival overall mechanisms are clearly shown, one is involving dissociation first followed by hydrogenation while the other one involves hydrogenation first followed by dissociation. It is quite likely that the $CH_3OH$ formation involves this latter mechanism. However, the formation of $NiCH_2$ the key intermediate in the Fischer-Tropsch synthesis may in principle proceed by either routes. Clearly, theoretical calculations, on the basis of molecular energetics, could decide which one of the two mechanisms is the more desirable one.

Acknowledgements

The author wishes to express his thanks and appreciation to Professor R. Daudel (Centre de Mécanique Ondulatoire Appliquée, CNRS, Paris, France) and Professor K. Morokuma (Institute for Molecular Science, Okazaki, Japan) for their hospitality. The financial support of the Natural Sciences and Engineering Research Council of Canada is gratefully acknowldged.

References

1. P. Sabatier and J.B. Senderens: (a) C.R. Acad. Sci. Paris, 134, 514 (1902) ; (b) J. Chem. Soc. Ind., 21, 504 (1902) ; (c) J. Chem. Soc., 88, 333, 401 (1905).
2. F. Fischer and H. Tropsch: (a) Brenstoff-Chem., 6, 265 (1925) ; (b) Brenstoff-Chem., 7, 97 (1926).
3. R. Pettit, private communication.
4. S.P. Walch and W.A. Goddard III, J. Am. Chem. Soc., 100, 1338 (1978).
5. K.P. Huber and G. Herzberg, Molecular Spectra and Molecular Structure, Vol. 4. Van Nostrand Reinhold Co. pp. 462-467.

ENVIRONMENTAL EFFECTS ON PROTON TRANSFER.
AB INITIO CALCULATIONS ON SYSTEMS IN A SEMI-CLASSICAL, POLARIZABLE ENVIRONMENT.

P.Th. van Duijnen and B.T. Thole
Theoretical Chemistry Group, Laboratory of Chemical Physics, University of Groningen, Nijenborgh 16, 9747 AG Groningen, The Netherlands.

*Some problems concerning the reaction mechanisms of papain and thiolsubtilisin are discussed. Results of ab initio SCF calculations, which contributed to the solution of those problems, are briefly reviewed. It is concluded that the environment of an enzyme's active site plays an important part in the activity.*

*In order to include environmental effects in quantum mechanical calculations, we recently proposed the "direct reaction field" method, in which environment is modelled by a collection of point-charges and -polarizabilities. The method, on theoretical grounds, should give a reasonably accurate estimate of interaction energies up to second order, in which dispersion is overestimated by a factor of about two. Actual calculations on the water dimer and a papain model system show that it is possible to simulate, with this method, ab initio calculations on large systems.*

## 1. INTRODUCTION: THE PAPAIN AND THIOLSUBTILISIN STORY

Papain is a proteolytic enzyme, the three-dimensional structure of which[1] shows the catalytic triad Cys25 - His159 - Asn175 in the active site. Drenth et al[2] proposed a catalytic mechanism for papain, based on the presence of a zwitterion Cys $S^-$ ...His$H^+$ in the free enzyme. This, in itself, is rather unexpected, since Cys $S^-$ should be a much stronger base than His. Normally, Cys and His have pK's of 9 and 6, respectively, in papain, however, one measures[3-6] 4 and 8 in the free enzyme, while His has pK ≈ 4 in papain with a blocked thiol group. This interpretation was not generally accepted[7-8] and further support could possibly be had from computations.

Thiol-subtilisin[9,10] is an artificial enzyme, obtained from native subtilisin by replacing the serine oxygen atom in the active site by sulfur, thus turning it into the triad Cys221 - His64 - Asp32. Although this looks very much like the active site of papain, and one would expect thiolsubtilisin to be a more or less normal enzyme, it is



*R. Daudel, A. Pullman, L. Salem, and A. Viellard (eds.), Quantum Theory of Chemical Reactions, Volume III, 85–95.*

not active at all towards proteins[11], and only very weakly towards specific substrates[12,13]. Again, most likely, computations could provide additional information.

A first series of ab initio calculations on models for papain[14] did not really solve the problem of the zwitter-ion, but gave an indication of the accuracy with which one has to perform such calculations. The discovery[15] of the strong electric field of the α-helix put us on the right track, i.e. that it is the *environment*, in which the active site residues are placed, which gives to them peculiar properties[16] and turns them into very efficient catalysts.

Similar arguments will be appropriate to other biochemical reactions, and, therefore, it is not only a matter of elegance that, in theoretical work on biomolecules, environmental effects should be taken into account properly. Since complete quantum mechanical treatment is beyond hope, it is not surprizing that there is a growing interest in reliable methods for taking such effects into account. Recently, Warshel and Levitt[17] and Tapia and Goscinski[18] developed methods which retain as much as possible of the structural details of the (complete) macromolecular system. Our "direct reaction field" approach[19] is, in fact, very similar to that of Tapia et al.[20] in that a (very) large system (solute/solvent or macromolecule) is split into a part treated quantum mechanically, and a part of which only the local field effect on this quantum mechanical part is considered. Our philosophy is that approximations should be introduced at appropriate points and in a balanced way. Moreover, the total description should be as "parameter-free" (or "ab initio") as possible.

In section 2 we will discuss - mainly for later reference - the general problem of intermolecular interactions. Section 3 contains a discussion of our papain calculations from the viewpoint of environmental effects. Section 4 gives an overview of our direct reaction field method with some numerical examples.

## 2. INTERMOLECULAR INTERACTIONS

Consider two (molecular) systems P and S with (local) hamiltonians $\hat{H}_P$ and $H_S$. Isolated, the wave functions are $\phi_P$ and $\phi_S$, respectively. Brought together, the hamiltonian becomes

$$\hat{H} = \hat{H}_P + \hat{H}_S + \hat{H}_{P,S} \tag{2.1}$$

and the wave function, for sufficiently good "separability"[21] between P and S, may be written

$$\phi = \hat{O}_A \, \phi_P \phi_S \tag{2.2}$$

where $\hat{O}_A$ is the appropriate antisymmetrizer. The interaction operator $H_{PS}$ can always be considered a one-electron operator on P or S. This can be achieved, either by assuming P and S sufficiently far apart, or by averaging $H_{PS}$ in a suitable way. The first possibility is obvious, the averaging taking place in either S or P, resulting in a potential generated by either system. The second possibility leads to the Hartree-Fock approximation (with P and S forming a supermolecule) or to McWeeny's group function approach[21]. Whatever the case may be, the total energy of the system can be written as

$$E = E_P^o + E_S^o + E' + E''_{IND} + E''_{DISP} \tag{2.3a}$$

with, in an obvious notation,

$$E_Q^o = \langle o|H_Q|o\rangle \tag{2.3b}$$

$$E' = \langle o|H_{P(S)}|o\rangle + \langle o|H_{(P)S}|o\rangle \tag{2.3c}$$

$$E''_{IND} = \sum_{p\neq o} \frac{\langle o|\langle o|H_{P(S)}|o\rangle|p\rangle^2}{\Delta_p} + \sum_{s\neq o} \frac{\langle o|\langle o|H_{(P)S}|o\rangle|s\rangle^2}{\Delta_s} \tag{2.3d}$$

$$E''_{DISP} = \sum_{s\ p\neq o} \frac{\langle o\ o|H_{SP}|SP\rangle^2}{\Delta_p+\Delta_s} \tag{2.3e}$$

where the brackets in $H_{(P)S}$ and $H_{P(S)o}$ indicate in which (sub)system the averaging takes place, and $\Delta_q = E_q^o - E_q$. The interpretation of 2.3 is simple. E' is just the interaction between two charge distributions in their respective fields. For overlapping charge distributions, E' also contains the exchange contributions. $E''_{IND}$ is similar, but after the two systems having polarized each other. These two terms (or rather their sum) are obtained automatically in any SCF procedure. Within a quantum mechanical appraoch, $E''_{DISP}$ can only be obtained from a second order perturbation of CI treatment of the complete system.

Formally, extension to any number of interacting (sub) systems is achieved by summing over all possible pairs of groups. Of course, in any practical case, full quantum mechanical treatment is out of bounds. In what follows, it is essential to realize that SCF procedures (supermolecule or group function approximation) exhausts the possible polarizations of the groups.

As will be shown in the following sections, $\hat{H}_{P(S)}$ and $\hat{H}_{(P)S}$ can be represented, for large enough separations between P and S, by suitable point-charge and point-polarizability models, so that E' and $E''_{IND}$ can be obtained in a semi-classical way, while for $E''_{DISP}$ an approximation is found.

## 3. AB INITIO CALCULATIONS IN A STATIC FIELD

Fig. 1 shows some models for the active site of papain. Methanethiol, imidazole and formaldehyde (formamide) represent Cys25, His159

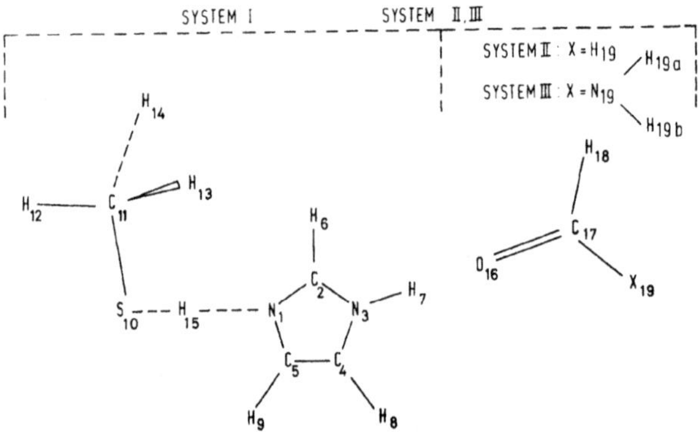


*Fig. 1 Model systems for the active site of papain.*

and Asn175, respectively. The question (see section 1) was, whether or not a zwitter-ion Cys $S^-$...His $H^+$ exists in free papain. Fig. 2 gives

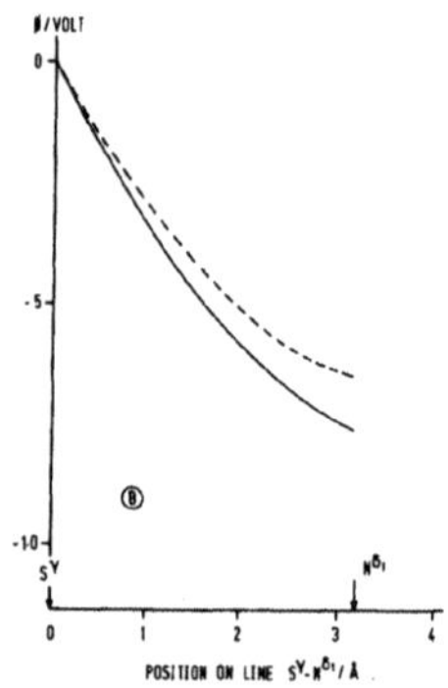

*Fig. 2 Electric potential generated by the α-helix in papain[3]. Solid line: ε=1, dashed line: ε=2.*

an impression of the static electric field, in the active site region, as generated by the α-helix[15] of which Cys25 is the second residue. Fig. 3 summarizes the results of *ab initio* SCF calculations[22,23] on the models of Fig. 1, with and without the helix, which was represented by a point-charge model of the backbone atoms[15].

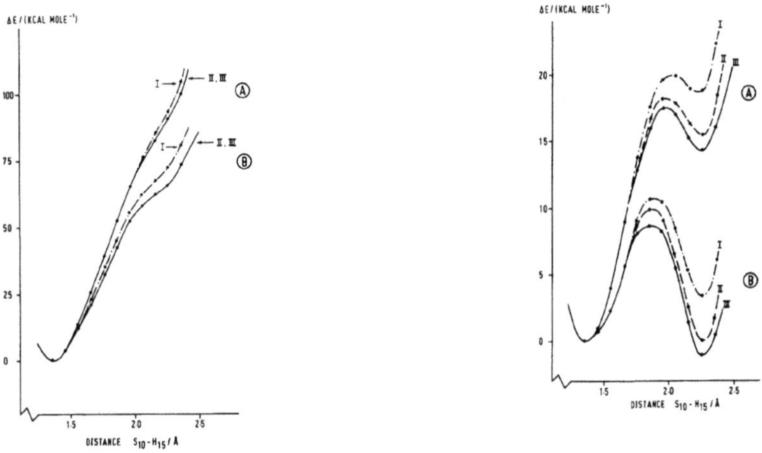


a b

*Fig. 3 Results of ab initio SCF calculations on the models I, II and III of Fig. 1. a. minimal basis set; b. double zeta basis set; A. without helix; B. with helix.*

Fig. 3b is almost self-explanatory: free papain most likely has the zwitter-ion configuration, due to the stabilizing effect of the helix field. Further analysis[23] shows, that this field is also responsible for the peculiar pK values (see section 1). In the case of thiolsubtilisin we expect a similar ion-pair configuration, which, due to the presence of negatively charged Asp⁻ instead of neutral Asn, will be stabilized by another 20 kcal/mol, or so[24,25]. The zwitter-ion in thiolsubtilisin was found, recently, by NMR techniques[26]. Hence, our calculations confirmed (papain) and predicted (thiolsubtilisin) the actual situation. The consequences for the reaction mechanisms are evident. In papain, after formation of a tetrahedral intermediate[27], His (with pK ≈ 4) will readily protonate the nitrogen atom of the scissile peptide bond. The latter is disrupted, thus leaving the acyl-enzyme. In thiolsubtilisin this will not happen, the pK of His-Asp⁻ being much larger than 4. Hence, no acyl-enzyme is formed, as was already noticed by the early investigators[11].

Of course, it is gratifying that theoretical work of this type leads to useful results for the experimental biochemist. For the theoretician, however, there is more relevant information in Fig. 3. First, comparing Figs. 3a and 3b, we see that minimal basis set calculations, or their equivalents, are inadequate for reproducing even qualitatively the behaviour of the systems at hand. Second, even in an extended basis, the experimental facts are only reproduced if the environment of the active site is represented adequately.

Obviously, for the latter aspect, the static field of the helix is dominating but not sufficient. Considering Cys-His as the active site, Asn belongs to its environment, and is essential, as can be seen from the differences between curves I, II and III in Fig. 3b. Similar stabilization of the ion-pair by other residues was found by Bolis et al.[28] In terms of the analysis of the preceding section, we see that the point-charge model for the helix introduces E' into the total energy. Since Cys-His-Asn, in systems II and III, are treated on the SCF level, the interaction between Cys-His and Asn also contains $E''_{IND}$. (Of course, also $E''_{IND}$ between the helix and Cys-His is included. Here, we want to emphasize the idea of Asn being part of the "environment"). The total energies in Fig. 3 for systems II and III, of course, also contain E' between the helix and Asn, but only part of $E''_{IND}$ between these fragments, since the helix, in this representation, is not polarizable.

In concluding this section, we note that
i) the ab initio quantum mechanical description of proton transfer at least needs a double zeta basis set for obtaining (semi) quantitative results.
ii) a classical description of the environment should account for static and polarization effects.

The next section describes a model which, to a large extent, meets these requirements.

## 4. THE DIRECT REACTION FIELD APPROACH

In our "direct reaction field" method[19] we try to represent the terms missing in $H_{P(S)}$ (see section 2) when the "environment" is modelled by point charges only. The philosophy of our method is basically the same as in other approaches[17,18], i.e. to the point charges we add (point) polarizabilities. The hamiltonian then contains the polarization energy of the total system as a function of the quantum mechanical part (P, say). Hence, the hamiltonian used in the ab initio calculations of the preceeding section is extended with terms

$$H^{RF} = -\tfrac{1}{2} \sum_{i,j,a,b} \underline{f}_{ia}^{\dagger} \underline{\underline{A}}_{ab} \underline{f}_{bj} \qquad (4.1)$$

where i,j sum over particles in $\underline{P}$ and a,b sum over the external points which constitute S. $\underline{f}_{ia}$ is the electric field of particle i (electron or nucleus) at the position of polarizable point $\underline{a}$. $\underline{\underline{A}}_{ab}$ is part of the polarizability or *transfer* matrix of S. It gives the dipole moment in $\underline{b}$ resulting from the applied field in $\underline{a}$. The elements of $\underline{\underline{A}}$ with a≠b arise from the *interactions* between induced dipoles.

Our method differs basically from other approaches, in that we introduce $H^{RF}$ *directly* into the hamiltonian of $\underline{P}$, prior to any decision as to how the wave function will be obtained. In our model, $\underline{\underline{A}}$ is constructed in a separate classical calculation on the atoms of S, using a modified dipole-dipole interaction[29]. The factor of ½ in (4.1) makes sure that the increase in self energy of S, due to the polarization by P, is accounted for.

For any type of wave function for P we have

$$E^{RF} = \langle H^{RF} \rangle = -\tfrac{1}{2}\langle \phi | \underline{F}^{\dagger} \underline{\underline{A}}\ \underline{F} | \phi \rangle \tag{4.2}$$

or

$$E^{RF} = -\tfrac{1}{2}\langle \underline{F} \rangle^{\dagger} \underline{\underline{A}} \langle \underline{F} \rangle - \tfrac{1}{2}\{ \langle \underline{F}^{\dagger} \underline{\underline{A}} \underline{F} \rangle - \langle \underline{F} \rangle^{\dagger} \underline{\underline{A}} \langle \underline{F} \rangle \} \tag{4.3}$$

The first term in (4.3) is clearly the induction term missing when S only contains charges. The second term, in curly brackets, by its form already, shows a dispersion like character. It is the expectation value of a quadratic form, minus the corresponding quadratic form of the expectation value of $\underline{F}$, i.e. it is the expectation value of the *fluctuations* in $\underline{F}$. In fact, as will be shown, this term gives an upper limit to the dispersion interaction. With the full spectrum of P we may use the closure relation:

$$\langle \underline{F}^{\dagger} \underline{\underline{A}} \underline{F} \rangle = \sum_{p=o}^{\infty} \langle o | \underline{F} | p \rangle^{\dagger} \underline{\underline{A}}\ \langle p | \underline{F} | o \rangle \tag{4.4a}$$

while

$$\langle \underline{F} \rangle = \langle o | \underline{F} | o \rangle \tag{4.4b}$$

Hence, we may write, for the second term in (4.3)

$$E^{RF}_{DISP} = \sum_{p \neq o}^{\infty} \langle o | \underline{F} | p \rangle^{\dagger} \underline{\underline{A}}\ \langle p | \underline{F} | o \rangle \tag{4.5}$$

Note, that the summation in (4.5) runs over *excited* states of P only. In the simplest case, where S contains only one polarizable point, we may write

$$\underline{\underline{A}} = \underline{\underline{\alpha}} = \sum_{S \neq o}^{\infty} \frac{|\langle o | R | s \rangle|^2}{\Delta_s} \tag{4.6}$$

and after averaging over angles and replacing $\underline{F}$ by a truncated multipole expansion, we have

$$E^{RF}_{DISP} = -\frac{2}{3} \frac{1}{R^6} \sum_{p,s \neq o} \frac{|\langle o | R | p \rangle|^2 |\langle o | R | s \rangle|^2}{\Delta_s} \tag{4.7}$$

R being the dipole operator.

Within the multipole expansion used, (4.7) and (2.3e) are identical, apart from the denominators: in (4.7) $\Delta_p$ is missing. This reflects the assumption, implicit in directly adding $H^{RF}$ to the hamiltonian of P, that the correlation between particles in P with those in S is instantaneous. In fact, S will react differently to particles (in P) with different velocities. The polarization of S will always lag behind the applied field, while, in contrast, the dipoles in the present model are always optimal in magnitude and orientation. It can be shown that this lag reduces the dispersion energy, relative to the completely classical approach, by a factor of about 4. This same factor is found in Buckingham's simple demonstration of dispersion effects[30]. Because in our model the response of P to fluctuations in the field from S is *not* taken into account, we expect our $E^{RF}_{DISP}$ to overestimate the actual interaction by a factor of about 2. Since $\underline{\underline{A}}$ can be taken to be a realistic molecular polarizability, in which $\Delta_s$ is properly included, we expect the induction energy, obtained form $H^{RF}$, to be fairly good.

For a spherical one-electron system we have

$$E^{RF}_{DISP} = -\tfrac{1}{2}\langle \underline{F}^{\dagger}\underline{\underline{\alpha}}\underline{F}\rangle = -\frac{\langle R^{\dagger}R\rangle\alpha}{R^6} = -\frac{3\langle x^2\rangle\alpha}{R^6} \qquad (4.8)$$

Two H-atoms would give $E^{RF}_{DISP} = -13.5/R^6$ where $E''_{DISP} = -6.4990/R^6$ [30]. Two harmonic oscillators would give $E^{RF}_{DISP} = -3/2\, \hbar^2/m^2\omega^2$ and $E''_{DISP} = -\tfrac{3}{4}\hbar^2/m^2\omega^2$. In these simple cases, the factor of 2 seems to hold quite well. As a further test, ab initio calculations were performed on the water dimer, comparing our model with the results of Derissen and Van Duijneveldt[31] who applied second order perturbation theory to the results of Matsuoka et al.[32]

Table 1 shows, that, indeed, the induction effect is treated quite well by our $H^{RF}$, while also the factor of about 2 seems to hold in more complecated systems.

*Table 1: Comparison of second order energies from SOP and direct reaction field method of the water dimer.*

| R(bohr) | $-E_{IND}$ | | $-E_{DISP}$ | |
|---|---|---|---|---|
| | DRF | SOP | DRF | SOP |
| 4.5 | 1.063 | 1.745 | 5.780 | 2.523 |
| 5.0 | 0.605 | 0.729 | 2.996 | 1.363 |
| 5.5 | 0.354 | 0.377 | 1.654 | 0.771 |
| 6.0 | 0.218 | 0.223 | 0.963 | 0.453 |
| 7.0 | 0.087 | 0.092 | 0.371 | 0.173 |
| 8.0 | 0.039 | 0.043 | 0.162 | 0.075 |
| 9.0 | 0.020 | 0.022 | 0.079 | 0.036 |
| 11.0 | 0.007 | 0.007 | 0.023 | 0.010 |

These results gave us sufficient confidence to try and recompute the proton transfer in System I (cf. Fig. 1) placed in an environment as found in actinidine[33]. This enzyme was chosen because structural data are more accurate than those of papain. In actinidine, the essential S-N distance is 3.0 Å, whereas in papain it was found to be 3.4 Å. Fig. 4 gives the energy profile for System II with an S-N distance of

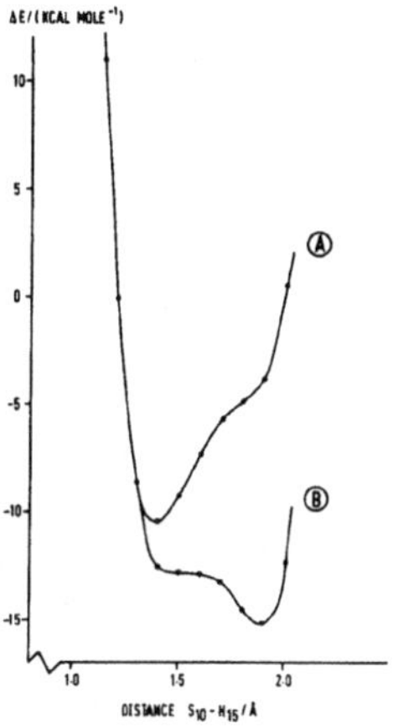


*Fig. 4 Result of ab initio SCF calculation for System II (cf Fig. 1) with S-N = 3.0 Å.*

3.0 Å[23]. Fig. 5 shows the present results for System I, i.e. without Asn. The energy profiles represent a step-wise introduction of the environment, starting with System I *in vacuo*, then adding the (point-charge) helix, then adding polarizable neutral atoms (except those of the helix) within a radius of 10 bohr from the midpoint between S and N, then, adding the atoms within a 30 bohr radius. Finally, an estimate, based on Fig. 3b, is given of the effect of including the charges of Asn in the last case.

Fig. 5 is in excellent agreement with Fig. 4. This shows, that it is possible to treat (very) large systems in a consistent way, which gives credit to molecular detail. A next step will be, to deal with the interactions within the "classical" part of the system more completely, i.e. using atoms that carry both charge and polarizability.

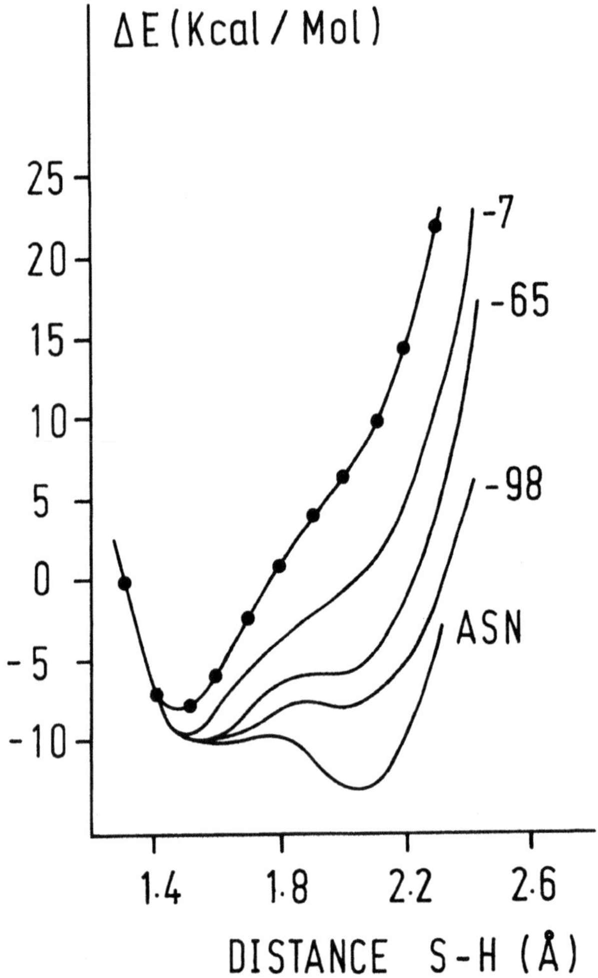


*Fig. 5 Results of ab initio SCF calculations on System I with stepwise introduction of the protein environment. From top to bottom:*
*System I in vacuo*
*point-charge helix added*
*polarizable atoms within 10 bohr added*
*" " " 30 bohr added*
*estimate, from Fig. 3, of the effect of taking into account the charges of ASN. Numbers in parentheses give the shift (kcal/mol) of the curves relative to the top curve.*

## References


1. J. Drenth, K.H. Kalk & H.M. Swen, Biochemistry 15 (1976) 3731.
2. J. Drenth, H.M. Swen, W. Hoogenstraaten & L.A.Æ. Sluyterman, Proc. K. Ned. Akad. Wet. C78 (1975) 104
3. L.A.Æ. Sluyterman and M.J.M. de Graaf, Biochem. Biophys. Acta 200 (1970) 595.
4. L. Polgar, Eur. J. Biochem. 33 (1973) 104.
5. L. Polgar, FEBS Lett. 38 (1974) 187; 47 (1974) 15.
6. G. Lowe and A.S. Withworth, Biochem. J. 141 (1974) 503.
7. G. Lowe, Tetrahedron 32 (1976) 291.
8. V.I. Zannis and J.F. Kirsch, Biochem 17 (1978) 2669.
9. K.E. Neet & D.E. Koshland Jr., Proc. Natl. Acad. Sci. US 273 (1966) 1606.
10. L. Polgar & M.L. Bender, J. Am. Chem. Soc. 88 (1966) 3153.
11. K.E. Neet, A. Nanci & D.E. Koshland Jr., Biol. Chem. 243 (1968) 6392.
12. I.H. Tsai & M.L. Bender, Biochem. 18 (1979) 3764.
13. M. Phillip, I.H. Tsai & M.L. Bender, Biochem. 18 (1979) 3769.
14. R. Broer, P.Th. van Duijnen & W.C. Nieuwpoort, Chem. Phys. Lett. 42 (1976) 525.
15. W.G.J. Hol, P.Th. van Duijnen & H.J.C. Berendsen, Nature 273 (1978) 443.
16. G. Johannin & N. Kellershow, Biochem. Biophys. Res. Commun. 49 (1972) 321.
17. A. Warshel & M. Levitt, J. Mol. Biol. 103 (1974) 227.
18. O. Tapia & O. Goscinski, Mol. Phys. 29 (1975) 1653.
19. B.T. Thole & P.Th. van Duijnen, Theor. Chim. Acta (Berlin) 55 (1980) 307.
20. See, for example, O. Tapia and G. Johannin, J. Chem. Phys. 75 (1981) 3624 and references cited there.
21. R. McWeeny & B.T. Sutcliffe "Methods of Molecular Quantum Mechanics, Academic Press, New York, 1969, ch. 5.
22. P.Th. van Duijnen, B.T. Thole & W.G.J. Hol, Biophys. Chem. 9 (1979) 273.
23. P.Th. van Duijnen, B.T. Thole, R. Broer, W.C. Nieuwpoort, Int. J. Quantum Chem. 17 (1980) 651.
24. P.Th. van Duijnen, Biophys. Chem. 13 (1981) 133.
25. P.A. Kollman & D.M. Hayes, J. Am. Chem. Soc. 103 (1981) 2955.
26. F. Jordan & L. Polgar, in "Steric effects in biomolecules" G. Naray-Szabo (Ed.), in press.
27. See for example A. Fersht, "Enzyme structure and mechanism", Freemen, Reading & S. Fransisco, 1977.
28. G. Bolis, E. Clementi, M. Ragazzi, D. Salvadero & D.R. Ferro, Int. J. Quantum Chem. 14 (1978) 815.
29. B.T. Thole, Chem. Phys. 59 (1981) 341.
30. A.D. Buckinham in "Intermolecular interaction: from Diatomics to Biopolymers (B. Pullman, Ed.) Wiley, NY 1978, Ch. I.
31. J.L. Derissen and F.B. van Duijneveldt, private communication.
32. O. Matsuoka, E. Clementi & M. Yoshimine, JCP 64 (1976) 1351.
33. E.N. Baker, J. Mol. Biol. 115 (1977) 263.

# RECENT QUANTUM/STATISTICAL MECHANICAL STUDIES ON ENZYME ACTIVITY[x] SERINE PROTEASES AND ALCOHOL DEHYDROGENASES.

O. Tapia, C.-I. Brändén and A.-M. Armbruster
Department of Chemistry and Molecular Biology
SLU, S-750 07 Uppsala, Sweden
C.M.O.A., 23 rue du Maroc, 759 40 Paris, Cedex 19, France

*Recent quantum and statistical mechanical studies on serine proteases and alcohol dehydrogenase are overviewed. Particular attention is given to the applications of the inhomogenous selfconsistent reaction field (ISCRF) theory of protein core effects. This theory allows for quantum calculations of active site systems immersed in the protein. The theory is described. Results on the protonation state of the catalytic triad in serine proteases are presented; ISCRF/INDO calculations of the acylation step are reported and comparisons are made with recent ab-initio studies. Several aspects of the catalytic mechanism of liver alcohol dehydrogenase are discussed. Particular emphasis is given to the hydride transfer step; Monte Carlo simulations of water structure in the substrate hydrophobic barrel and active site are reported. The nummerical results obtained with the ISCRF scheme show the fundamental role played by the protein core potential in describing enzyme catalytic properties.*

## 1. INTRODUCTION

Theoretical studies of enzyme catalysed reactions have been abetted by rapid progress in high resolution protein structure determinations. Accurate structural data on enzyme active site stereochemistry and complexes with substrate analogs have been derived from diffraction methods. Model building techniques have helped setting up intermediate structures demanded by the reaction mechanisms, that are compatible with the protein structure and general organic chemical knowledge. Such detailed structural information on reaction pathways has prompted a host of quantum chemical studies.[(1-9)]

Furthermore, significant strides have been made on developing appropriate theoretical frameworks to study enzyme reactivity. Thus, we have seen further extensions of the electronic theories which permit calculating active site molecular model systems in the presence of the surrounding protein and eventually the solvent.[(5-8)] These theories open a way to study the effects of the protein structure and/or solvent



*R. Daudel, A. Pullman, L. Salem, and A. Viellard (eds.), Quantum Theory of Chemical Reactions, Volume III, 97–123.*

changes on the active site reactants. In quantum chemistry, methods and programs have been worked out in view of calculating electronic wave functions for biological significant systems. In statistical mechanics, simulations of solvent structure in bulk and solvation effects on bio-molecules are becoming common tools; Monte Carlo and molecular dynamics simulation of protein and macromolecules solvation have been reported.[9,10]

In this paper we summarize the results obtained with the inhomogenous selfconsistent reaction field theory of protein core effects.[7,8] The theory has been applied to study serine proteases and alcohol dehydrogenases. Recent studies done by other authors relevant to the catalytic mechanism of these enzymes are incorporated in the discussion of our own results.

For the sake of completeness in the following section an outline of the ISCRF theory is presented. In section 3 serine proteases are examined; First, the most relevant biochemical and structural facts are presented; Second, the quantum chemistry of the catalytic mechanism is discussed. In section 4 a similar pattern is followed to examine the results on alcohol dehydrogenases. A general discussion is given in the last section.

x) *This work has been partially suported by the Swedish Natural Science Research Council through contracts no K-UR 4575-100 and R-RA 4575-102. Cyber 170 computer time was granted by Uppsala Data Central through the project ULT10.-139010.*

## 2. THEORETIC OUTLINE

A proper description of catalytic sources at the electronic level requires a quantum mechanical framework. With the development of computer technology and software, the number of quantum chemical studies on molecular models of enzyme mechanistic steps, as well as on properties and mechanisms of biologically active molecules,[2] has been steadily growing. There, in vacuo supermolecule approaches are commonly in use. While studies with this model are commonly indicative of the nature, source and type of relevant protein-substrate (and drug-receptor) interactions, there are several features which can not be described with these standard quantum chemical calculations, namely temperature and surrounding medium dependence. In particular, many catalytic events are underpinned by spatial charge separation processes via electron and/or proton transfers. These processes play central roles in enzyme activity and photobiology. In these cases the surrounding medium effects can be metaphorically described as filtering and/or amplifying particular potentialities of the system in vacuo, which, otherwise would have never been reached. On the other hand, temperature may become a controlling factor by affecting the folding state and solvent structure around the active site system. Therefore, inclusion of surrounding medium effects into quantum chemical framework becomes a compelling requirement.

Recently, we have derived an inhomogenous selfconsistent reaction field theory which enables the calculation of a given reactive subsystem inside or at the surface of a globular protein. A summary of it follows.

### 2.1. Basic Model

The model underlying the study of the electronic structure of a reactant (or just interacting) subsystem in an enzyme is as follows: First, an appropriate cutout volume, v', inside or at the surface of the protein structure is defined; Second, a model molecular system is built in the volume v', which includes relevant functional groups of the protein, substrate, coenzyme (if any) that take part in the molecular mechanism; This system will be referred to as either the electroactive, protonactive or cage system. In this model bonds to atoms outside the cage boundaries are saturated with simple groups like hydrogen atoms, methyl or ethyl groups. The remaining atoms of the protein structure outside the cage, including tightly bound solvent molecules, cations or anions, provides local discrete environments to the cage system. This ensemble is referred to as the protein core. The effects of the environment of the protein such as bulk solvent, biomembrane etc. are left out of our present considerations.

Two sources contribute to the interaction hamiltonian operator, $H_{INT}$, which describes the coupling between the electroactive system and the protein core. They are represented by:

a) a total charge distribution model for the protein core
b) an equivalent charge polarization density

The preceding charge densities are represented by $P_{PC}(\vec{q})$ and $P_{POL}(\vec{q})$, respectively. The effective hamiltonian for the cage molecular system can be cast into the simple form:[7,8]

$$H = H_o + \int_V d\vec{r} \rho_c(\vec{r}) \int_{\bar{V}} d\vec{q} \{P_{PC}(\vec{q}) + P_{POL}(\vec{q})\} T(\vec{r},\vec{q}) \quad (1)$$

where V is the total volume of the system; $\bar{V} = V - v'$; $\rho_C(\vec{r})$ is the total charge density operator of the electroactive system; $T(\vec{r},\vec{q})$ is a shorthand notation for the Coulomb operator: $T(\vec{r},\vec{q}) = 1/|\vec{r}-\vec{q}|$; $H_o$ denotes the hamiltonian of the model cage system as if it were in vacuo.

The equivalent charge polarization density is set up by the field $\vec{D}(\vec{r})$, the displacement field. The field's source is the quantum mechanically averaged cage potential gradient:

$$\vec{D}(\vec{q}) = -\vec{\nabla}_q \langle \Psi | \int d\vec{r} \rho_c(\vec{r}) T(\vec{q},\vec{r}) | \Psi \rangle \quad (2)$$

Here, $\psi$ is the n-electron wavefunction for the cage system in equilibrium with the protein core; the brackets indicate standard quantum mechanical average over the spin and spatial coordinates of the electrons and nuclei; $\vec{\nabla}_q$ is the Laplacian vector operator with respect to q-coordinates.

The material support for the charge polarization density is provided by the polarizable matter in the protein core and beyond. Let us represent it by $\bar{\bar{\chi}}(q)$, i.e. the dielectric susceptibility tensor of the protein core. This tensor characterizes the response of the protein matter to the applied field $\vec{D}(\vec{q})$. The response is measured by the induced dipole density:

$$\vec{m}(\vec{q}) = \bar{\bar{\chi}}(\vec{q}) \cdot \{ \vec{D}(\vec{q}) + \int d\vec{q}' \bar{\bar{T}}(\vec{q},\vec{q}') \cdot \vec{m}(\vec{q}') \} \quad (3)$$

here $\bar{\bar{T}}$ is the instantaneous dipolar tensor propagator:

$$\bar{\bar{T}}(\vec{q},\vec{q}') = \vec{\nabla}_q \vec{\nabla}_q T(\vec{q},\vec{q}') \quad (4)$$

Introducing concepts of classical electrodynamics, the polarization density can be cast into

$$P_{POL}(\vec{q}) = -\int d\vec{q}' \vec{m}(\vec{q}') \cdot \vec{\nabla}_q \delta(\vec{q}-\vec{q}') \quad (5)$$

where $\delta(\vec{x})$ is the Dirac functional.

Thus, once a model for the protein charge distribution is chosen the effective hamiltonian is fully defined. Its building blocks are the three dimensional structure, and the dielectric properties of the protein. As the effective hamiltonian is known, the problem is to find out the energy and wave function for the cage system as a function of external (surrounding) and internal geometric parameters.

To get the solution of interest, one first searches for solutions for the model system in vacuo, i.e., a supermolecule calculation with $H_o$. Thereafter, the wave function for the active site model in the protein is calculated by solving the effective Schrödinger equation:

$$H|\Psi\rangle = \{H_o + H_1(\Psi)\}|\Psi\rangle = E_{eff}|\Psi\rangle \tag{6}$$

in a selfconsistent fashion. $E_{eff}$ is then the energy of the cage system in equilibrium with its surroundings.

Notice that eq. (6) depends on the moments of the cage charge density via eqs. (5), (3) and (2). This leads to a non-linear Schrödinger equation[8]. Consequently, eq. (6) has to be solved selfconsistently, otherwise important effects might be lost.

From the physical point of view it is important to obtain reasonable solutions of the Schrödinger equation in vacuo; i.e.,

$$H_o|\Phi_k\rangle = E_k^o|\Phi_k\rangle \tag{7}$$

The solutions of this equation may signal out potential sources of reactivity changes that may take place in substrates interacting with receptors in the active site. Furthermore, protein core effects may be rationalized in terms of a reshuffling of the energy levels of $H_o$. For a number of important cases, cross-over in $H_o$ energy states are found as an outcome of surrounding medium effects. Experimental evidence exists to substantiate such cross-over effects.

In résumé, as highly polarized states of $H_o$ are bound to be substantially shifted by protein effects, the ground state properties of $H(\Psi)$ might consequently be profoundly changed when a comparison is made with the system in vacuo. This is the rationale to search after selfconsistent solutions to eq. (6).

2.2 Calculational methods

For practical calculations we have put together the techniques currently employed in protein physical chemistry and those used in computational quantum chemistry. No attempt has been made to adjust any parameter in these procedures.

For the electronic calculations we have retained approximate molecular orbital schemes: INDO/CNDO-framework.[11] Special subroutines have been written to incorporate the ISCRF matrix elements.

From protein physical chemistry we have retained the point charge approximation to the total charge density:

$$P_{PC}(\vec{q}) = \sum_{a=1}^{N} \xi_a \, \delta(\vec{q}-\vec{q}_a) \tag{8}$$

where $\xi_a$ is the Warme-Scheraga set of point charges for the atoms and residues in the protein;[12] $\vec{q}_a$ denotes the position vectors of such atoms; N is its total number.

Molecular polarization data has been used to calculate atom and group polarizabilities $\alpha$.[12] To within the same level of approximation as in eq. (8), the dielectric susceptibility is calculated with:

$$\bar{\bar{\chi}}(\vec{q}) = \sum_{a=1}^{N} \bar{\bar{\alpha}}_a \; \delta(\vec{q}-\vec{q}_a) \tag{9}$$

The coordinates defining the set of $q_a$ are obtained from the X-ray model when available. Monte Carlo simulations and model building techniques are used to obtain coordinates for additional water molecules and bound substrate molecules, respectively.

For further details and derivation the reader is referred to our published papers.[7,8] Let us turn the attention to discussing some of the applications made on serine proteases.

## 3. SERINE PROTEASES

The catalytic mechanism of serine proteases has been the subject of extensive and intensive biochemical, structural and theoretical work.[13-27] A number of puzzling features remain to be correctly answered at the electronic molecular level. Any theory aiming at describing enzyme catalytic properties has to be confronted with available experimental knowledge of these enzymes to test whether or not sensible answers are given to the issues raised by their mechanism. Thus, after recalling some basic biochemical and structural facts concerning them, the results obtained with our approach are presented. A discussion of related work by other authors is incorporated here.

### 3.1. Structure-function relations

Serine proteases catalyse the hydrophobic cleavage of peptide and ester bonds. Two structurally unrelated families occur: The trypsin and subtilisin types. Chymotrypsin, trypsin, elastase belong to the former type, and subtilisin to the latter.

Subtilisin and trypsine type enzymes have their catalytically functional groups arranged in the same geometrical relationship although they are structurally unrelated. The presence at the active site of a "catalytic triad" comprised of the side chain residues of serine (Ser) histidine (His) and Aspartate (Asp) have been demonstrated for all of them.[13]

The serine side chain forms a transient covalent linkage with the carbonyl carbon atom in the susceptible ester or amide bond of substrate molecules. Since uncatalysed alcoholysis of peptide bonds is thermodynamically favourable, albeit kinetically slow, formation of the acyl intermediate is not in itself helpful in understanding why the enzymatic hydrolysis is so effective. The presence of Aspartate and Histidine side chains, in a rather invariant spatial arrangement with respect to the reactive serine, has been taken as a proof that they participate in the catalytic activity.

Ever since Blow, Birktoft and Hartley proposed a "charge relay mechanism", the actual protonation state in the triad at different steps of the mechanism has been a source of controversy and inspiration for theoreticians and experimentalists alike.[3,4,13-27]

Schematically, the Asp side chain is the general base during the formation of the tetrahedral intermediate in the charge relay mechanism, while in the proton relay mechanism[14] the general base is the imidazole side chain. In the former case the canonical triad:

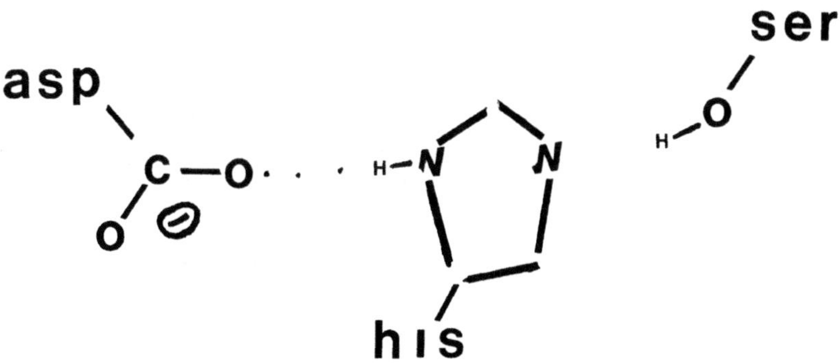


I

undergoes a two-proton shift to get:

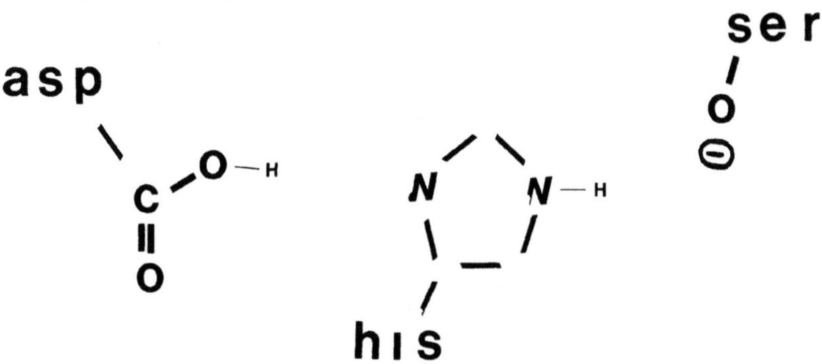


II

which will be referred to as the charge relay structure (CRS); while in the latter case one gets:

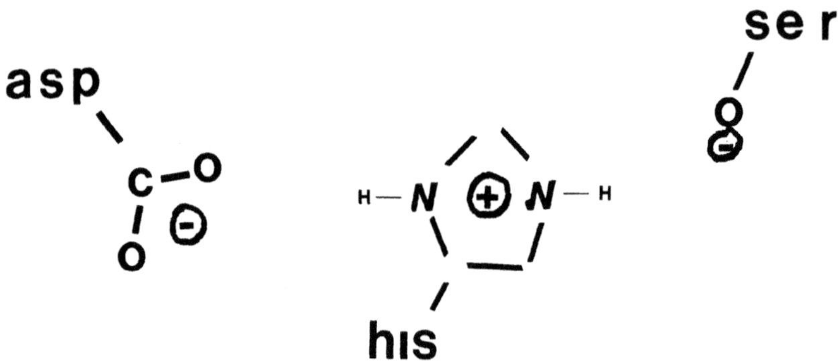


III

which will be referred to as the ion-pair form or zwitterionic structure (ZS).

The vital role played by histidine in the catalysis by serine proteases has been well established.[16] The activity of these proteases decreases with decreasing pH down to the pH where the imidazole is supposed to be converted to its conjugate acid. This has been taken as evidence that the unprotonated imidazole ring is the active form. Recently, a study of stereochemical aspects of peptide hydrolysis on the basis of extensive structural data has led to the following mechanistic role for histidine:[20] a conformational change allows the imidazole ring of His-57 to occupy two distinct positions; in one position a proton is abstracted from OG of Ser-195, and in the other this proton is donated to the leaving nitrogen.[20] Neutron structure analysis at 2.2 Å resolution has conclusively shown that the catalytic base in the transition state of the reaction is His-57, not Asp-102.[19]

A re-examination of the catalytic triad structure with high resolution data for subtilisin and comparisons with trypsin family members showed in 1977 some surprising results concerning the interactions between serine-histidine side chains on the one hand, and histidine-aspartate side chains on the other hand.[14] There seems to be no H-bond between Ser-221 and His-64 in native subtilisin, or at least that any such H-bond is severely distorted. The high resolution structures of trypsin, α-chymotrypsin, α-lytic proteases, SGPA and elastase confirm the subtilisin results. In contrast, for histidine-aspartate there is a rather strong H-bond. Their geometries, as observed in the trypsin family, do not show evidence of bifurcation between the two carboxylate oxygen atoms. A similar result was obtained for subtilisin. It can be concluded from these extensive structural data that His-Asp H-bond in all cases is a strong bond with normal (almost linear) geometry.

We have given a great deal of attention to experimental facts. They tend to show a rather clear picture of some steps in the proposed mechanism for these enzymes. The discussion, however, can not be exhaustive. Further details and qualifications can be found in the pertinent references cited here. The information quoted is enough to discuss the theoretical works presented in the following section.

3.2 Quantum Chemical Studies

One and two proton transfer steps of the catalytic mechanism of serine proteases have been subjected to a number of quantum chemical studies. Early works have already been reviewed.[4,8] The attention is hence focused on recent ab-initio and approximate molecular orbital calculations.

3.2.1. Catalytic triad without substrate. A typical two proton process is the passage from structure I to structure II, while the attainement of III from I corresponds to a one proton transfer step. There is a second one-proton transfer process corresponding to the transition from I to IV:

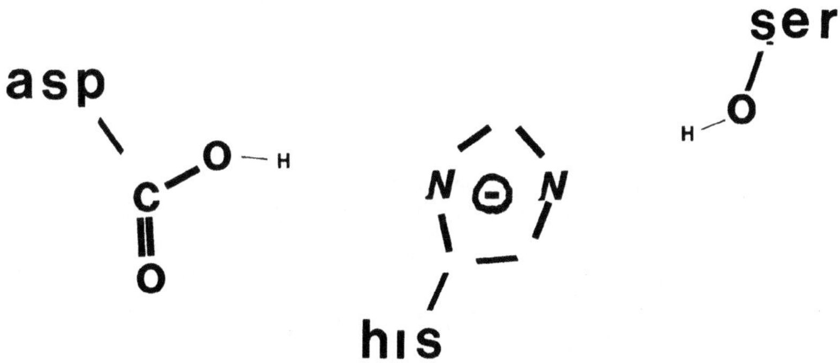


IV

Although structure IV has never been proposed in any mechanism of serine proteases, it is a structure that cannot be ruled out in a quantum chemical study. The reason is simple if one compares the proton affinities (PA) involved in these one proton steps: $PA(Im^-)$ = 354; $PA(HCOO^-)$ = 342, PA(Im) = 220 and $PA(CH_3O^-)$ = 377 kcal/mol. In vacuo, the attainement of III (neglecting electrostatic interactions) requires more energy than IV.

INDO-ISCRF calculations and α-chymotrypsin, trypsin, elastase and subtilisin have been carried out.[28] After the equilibrium proton positions are determined and the geometry of the protonated carboxylate reoptimized, the calculations are made on the four structures by steps:

i) the supermolecule in vacuo
ii) the supermolecule subjected to the protein electrostatic potential
iii) the supermolecule coupled to the protein core field.

Typical results are illustrated in table 1 for elastase and chymotrypsin.

Table 1

| | | | | |
|---|---|---|---|---|
| | Elastase | | | |
| Supermolecule | I | II | III | IV |
| In vacuo | 0 | 6.7 | 26.1 | 11.8 |
| Protein permanent field. | 0 | 8.3 | 3.9 | 52.9 |
| Protein core field | 0 | 8.3 | 24.7 | 69.8 |
| | Chymotrypsin | | | |
| In vacuo | 0 | 9.6 | 26.2 | 5.7 |
| Protein permanent field | 0 | 9.2 | 56.2 | 24.7 |
| Protein core field | 0 | 7.9 | 67.0 | 39.3 |

*Energy differences (in kcal/mol) between the canonical structure (I) and the charge relay (II), zwitterionic (III) and imidazolate (IV) forms. The entries have been corrected for proton affinity errors following Hayes and Kollman procedure* [25,29]. *Elastase and chymotrypsin coordinates are taken from references (30) and (31), respectively. The position of serine oxygen and orientation of the histidine ring in chymotrypsin correspond to the native form.*

One notice first that INDO results for the systems in vacuo are consistent with earlier approximate molecular orbital studies.[4] Recently, a calculation with PCILO method[24] has shown the carboxylate group of the aspartic acid residue to be the ultimate proton acceptor in the catalytic triad of a model chymotrypsin. The entries in table 1 show the same trend; the results for subtilisin and trypsin are found to be in between those reported here. The zitterionic form has too large energy to become a populated state. These results are at variance with ab-initio (4-31G)[32] molecular orbital calculations on β-trypsin.[23] 4-31G results show that the zwitterionic structure is below the charge relay form by about 13 kcal/mol; structure III is above the canonical one by 24.6 kcal/mol. These latter results are not corrected by proton affinity errors, that are present in these calculations as well. Albeit corrected figures do not change the order. Unfortunately, in this

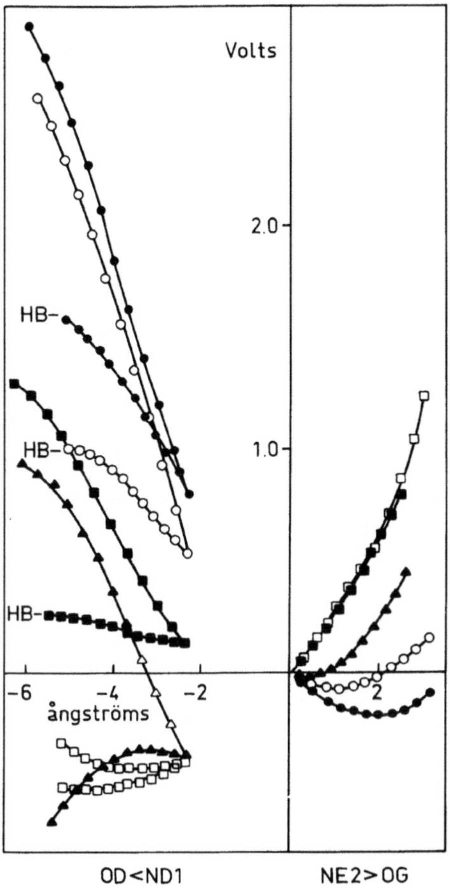


Figure 1. ELECTROSTATIC POTENTIAL FOR SERINE PROTEASES.

The curves are referred to the potential at NE2 which is arbitrarily given a zero value. The exact position of the curves can be obtained by a "translation" of :

| | | |
|---|---|---|
| • • • • • • | Chymotrypsin | 2.994 volts |
| ○ ○ ○ ○ ○ ○ | Trypsin | 7.495 volts |
| ■ ■ ■ ■ ■ ■ | Elastase | 6.560 volts |
| ⌑ ⌑ ⌑ | Subtilisin | 3.262 volts |
| ▲ ▲ ▲ ▲ ▲ | S G P A | 5.317 volts |

No screening dielectric effect is introduced at this stage.

(Fig. taken from G. Johannin, CECAM Workshop on Enzyme Catalysis Simulation Paris, Orsay, 1979)

ab-initio work[23] there is no indication of the geometries used for the protonated and unprotonated carboxylate. Furthermore, no minima were found in the two-proton 4-31G energy hypersurface for the charge relay and the zwitterionic forms. We have made calculations with a non relaxed protonated carboxylate for the C12 form. The result is to produce a large destabilization of the CR with respect to the zwitterionic species.

As expected, structure IV is definitely more stable than the zwitterionic form. What is surprising is that the imidazolate intermediate is more stable than the charge relay in chymotrypsin, and it is only 5 kcal/mol over II for elastase. 4-31G results for β-trypsin[23] display the same trend. Quantum chemical results assign an important role to this one proton step for the systems in vacuo. This is certainly in contradiction with current biochemical experience. Ab-initio and approximate molecular orbital calculations of such systems in vacuo are certainly inadequate to describe enzymatic spatial charge transfer processes.

ISCRF studies show that it is the protein surrounding field which put a huge energy barrier to the imidazolate structure. 4-31G studies where the field of some neighbouring residues are included show this effect as evidenced from table 1.

The electrostatic protein potential stabilizes the zwitterionic structure of elastase and produces the opposite effect for chymotrypsin. To rationalize the protein permanent field effects we reproduce in fig.1 the potential neated by $P_P$ along the axis between OG(Ser) and NEZ(His) and ND1(His) and OD2(Asp), and ND1(His)-OD1(Asp). A glance at this figure shows that the quantum results given in Table 1 elicit the changes in the potential between serine-histidine-aspartate. However, the charge translocation from the carboxylate to the serine is always favoured by the classical electrostatic potential, while for elastase the quantum results show up a destabilization. The classical electrostatics and quantum mechanical outcomes for the zwitterionic form coincide. It can also be seen that the displacement of a negative charge from the carboxylate towards imidazole is always made against the protein field which oposes it. The classical electrostatic potential maps are then quite useful to hint protein effects on charge displacement processes, however they cannot supplant the actual quantum mechanical study.

The protein electrostatic and polarization field contribute to destabilize the imidazolate beyond energetically accessible states during a chemical reation of this type. Surprisingly, the polarization field also contributes to a strong destabilization of the zwitterionic form. Fig. 1 shows that the protein permanent field is roughly directed from ND1 towards NE2, while formation of the ion-pair produces a huge dipole moment with a component having the opposite direction. As the polarization field tends to increase this dipole moment, the overall zwitterionic form becomes strongly destabilized by the permanent protein field.

The carboxylate oxygen atom which is H-bonded to imidazole has the smallest electrostatic potential. One can conjecture then that the negative charge on this group will tend to be trapped by this region of high potential; This effect might be enhanced by hydrogen bonding of OD to surrounding cage groups.

One can rather safely conclude that for the "catalytic triad" without substrate the zwitterionic form postulated in the presently accepted mechanism is not energetically accessible. It is therefore necessary to examine the behaviour of the different protonation states in the presence of a model substrate.

3.2.2. Catalytic triad with substrate. Methyl acetate interacting with the catalytic triad has been studied with the INDO-ISCRF calculational scheme.[23] Model building techniques were used to obtain substrate coordinates for each intermolecular distance R; (This distance is measured between OG(Serine) and the carboxylate carbon atom); Substrate-protein distances for each R-value were optimized by rotating the substrate as a rigid body. The substrate is further constrained so as to maximize the contacts between the carbonyl oxygen atom of the substrate and the N-H peptide groups from Ser 195 and Gly 193; (The so called "oxyanion hole"; in fact, the electrostatic potential in elastase at the position occupied by this atom is 7.1 V at R=2.0 Å, while at the atoms of the ester carboxylic group there is a potential of ca 6V). A tetrahedral deformation is introduced in the substrate to follow energy changes during formation of tetrahedral intermediates.

We notice, of course, that such a small substrate cannot incorporate interactions with primary and secondary binding sites of the protein. These interactions, govern the intermolecular distance R in the tetrahedral intermediate besides determining substrate-enzyme specificity. The INDO-ISCRF calculations describe the changes produced by the protein on the electronic structure of the susceptible amide or ester group.

Numerical results for elastase are presented in tables 2a and 2b.

Table 2a

| Distance (Å): | 4.0 | 2.0 | 1.5 |
|---|---|---|---|
| CF(I) | 0 | - 10.7 | 31.1 |
| CR(II) | 10.2 | - 43.0 | - 79.7 |
| ZW(III) | 26.0 | - 34.7 | - 78.0 |

*Potential energy as a function of distance between the substrate and the catalytic triad in vacuo.*

Table 2b

| Distance (Å): | 4.0 | 2.0 | 1.5 |
|---|---|---|---|
| CF(I) | 0 | - 36.8 | 2.2 |
| CR(II) | 55.8 | - 37.7 | - 81.4 |
| ZW(III) | 14.5 | - 79.3 | - 126.2 |

*Potential energy in presence of the protein core field.*

Consider first the results for the supermolecule in vacuo. The canonical structure (I) as a function of the substrate-triad distance shows a minimum in the potential energy curve. In contrast, for the charge relay and zwitterionic forms the energy decreases with decreasing intermolecular distance. At R=4.0 Å the canonical form is more unstable. At shorter distances there is an inversion of stability; at 2.0 Å the charge relay form is the most stable form. At shorter distances the catalytic triad has almost equal probabilities to populate the charge relay and the zwitterionic forms.

Comparing the preceding results with the catalytic triad without substrate, one can see that the substrate-triad interaction is responsible for a huge stabilization of the zwitterionic form. However, it does not show a clear distinction between the two mechanisms. Assuming that the tetrahedral intermediate would be formed at around R=2.0 Å these calculations show the charge relay is the favoured mechanism, if the supermolecule is treated in vacuo.

A more clear picture comes out once the protein core effects are incorporated in the quantum scheme (c.f. table 2b). The preceding trend for the system in vacuo is dramatically accentuated. The zwitterionic structure at 2 Å and beyond is bound to be the dominant form.

At 4 Å the formation of a tetrahedral intermediate requires 20 kcl/mol for the zwitterionic form, and 22 kcal/mol for the triad in the canonical form. At 2 Å the same deformation is fully exothermic.

Therefore, a small barrier ought to be found along the catalytic path at the intersection between the curves corresponding to the canonical and zwitterionic forms. The charge relay structure is being prevented by the protein core field.

It is interesting to note that the foregoing results are in agreement with recent ab-initio 4-31G calculations by Allen[(33)]. Polarization effects were included via a simple dielectric constant factor to screen the protein electrostatic potential.

It can thus be concluded that INDO-ISCRF representation scheme of supermolecules in protein cages does describe spatial change transfer processes in a rather reliable manner.

## 4. ALCOHOL DEHYDROGENASES

Studies of dehydrogenases have been important in the development of some general concepts in enzymology. The structure and mechanism of liver alcohol dehydrogenase,[34] lactate dehydrogenase[35] and glyceraldehyde-3-phosphate dehydrogenase[36] have been thoroughly analysed. The reactions catalyzed by these enzymes are apparently very simple reactions. In the simplest form an alcohol group of a metabolite is oxidized by the removal of a proton from the hydroxyl group and the transfer of a hydride ion from C-1, the carbon atom that binds the hydroxyl group, to the oxidized form of the coenzyme. A second reaction is sometimes coupled to the dehydrogenation such as phosphorylation in the case of GAPDH and decarboxylation in the case of 6-phosphogluconate dehydrogenase. An essential point for a discussion of the structural requirements for these reactions is that a hydride ion is directly transferred from C-1 of the substrate to the C-4 atom in the nicotinamide moiety of $NAD^+$.[37]

The overall reaction is, however, complicated by a number of structural requisites. These enzymes must recognize and bind NAD in such a way that the nicotinamide is correctly positioned in the active site. The binding energy of NAD to an enzyme is potentially very high. Fersht[38] estimates that a dissociation constant of $10^{-20}$ M could occur if these potential binding energies were realized. Since reasonable high $K_m$ values are an important component of efficient catalysis it is thus essential that some of the possible interactions between enzyme and coenzyme do not occur. Thus there is a conflict between the requirements of recognizing NAD and positioning the nicotinamide on the one hand and sufficiently weak binding for efficient catalysis on the other. This conflict may be one of the important reasons why the dehydrogenases are relatively slow enzymes, frequently with the dissociation of coenzyme as the rate limiting step.

Furthermore, these enzymes must recognize and bind the substrates with various degrees of specificity in such way that C-1 is correctly positioned and oriented in the active site. However, the substrate molecule and the nicotinamide moiety may not be rigidly fixed relative to each other. The electronic rearrangement of C-1 from $sp^2$-hybridization in the aldehyde to $sp^3$ in the alcohol requires different preferred pathways for hydride transfer and thus different relative orientations. Thus it is generally found in these structures that the nicotinamide moiety is not rigidly fixed but positioned in a hydrophobic cavity which is sufficiently large to allow different orientations of the ring.

Dutler[39] had described such possible movement for alcohol dehydrogenase in terms of a wind shield wiper arrangement.

Direct transfer of a hydride ion is facilitated in a hydrophobic surrounding where water is excluded. Thus there is also a conflict between this requirement and the fact that the active site must be open and available for positioning of both nicotinamide and substrate. This apparent conflict is resolved by induced conformational changes of the protein structures such that the active site is shielded from the solution after binding of coenzyme and substrate.

Quantum chemical studies on the catalytic mechanism of these enzymes are rather scarce. A supermolecule in vacuo calculation of lactate dehydrogenase charge realy system has been reported.[7,29] We have studied liver alcohol dehydrogenase (LADH) proton relay system with the ISCRF theory.[7] Furthermore, several quantum chemical studies on molecules and processes having a bearing on these enzymes have been published.

In the following attention is focussed on LADH (principally). Some structure-function relationships are first recalled. Then an overview of quantum studies partially related to the catalytic properties of LADH is presented. There particular attention is given to zinc-ligand interactions and the electronic structure of the hydride transfer step in simple model compounds. Finally descriptions of our work on the proton relay mechanism and the statistical mechanical Monte Carlo simulation studies on water structure in the active site and substrate barrel of LADH are presented.

4.1 Structure-function relationships: LADH.

LADH is a zinc containing and nicotinamide (AD) requiring enzyme.[34,37] It catalyses the oxidation of variuos primary and secondary alcohols to aldehyde and ketones, respectively.

The molecule is formed by two chemically identical subunits. Each subunit is divided into two domains. One of them binds the coenzyme NAD, the other provides the residues that bind two zinc atoms. The zinc atoms are not equivalent, one has a catalytic role, the other a structural role. The catalytic zinc atom is liganded to three protein atoms: The N atom of His 67 and the sulphur atoms of Cys-46 and Cys-174. In the apoenzyme the coordination sphere of zinc also contains a water molecule. This catalytic zinc is found at the bottom of a deep pocket in a region between the domains forming each subunit.[34,37]

The enzyme has an ordered mechanism. The binding of oxidized coenzyme ($NAD^+$) precedes productive binding of alcohol to the protein. NAD binding triggers a conformational change, and a substantial pKa shift of the water molecule bound to zinc is detected. The isomerization to the holoenzyme structure is facilitated in the presence of alcohol.

The alcohol is oxidized via a proton translocation towards the hydroxyl ion bound to zinc with formation of an alcoholate ion bound to the catalytic zinc. The pK values for alcohols are drastically decreased when bound to zinc.[44] In a subsequent step a transfer of an hydride ion from C-1, the carbon atom that binds the alcohol group, to the C-4 atom in the nicotinamide moiety of $NAD^+$ is acomplished.

As it was pointed out, direct transfer of an hydride ion is facilitated in an hydrophobic surrounding where water has been excluded. This conflicts with the fact that the active site must be open and available for positioning of both NAD and the substrate. A sort of compromise is reached: The conformational change induced by coenzyme binding closes the entrance to the active site from the front of the molecule, but still leaves space to evacuate the water molecules. Substrate uptake completes this process of hydrophobization.

For the pKa shift during coenzyme binding, a proton relay mechanism has been proposed. Water bound to the catalytic zinc, serine 48 and histidine 51 could form a network connecting the active site with the surface of the protein. The protein would be activated for substrate binding by creating an hydroxyl ion bound to zinc when the cationic nicotinamide ring comes in to the neighbourhood of the catalytic zinc.

Compared to the mechanism of serine proteases,those of LADH and other dehydrogenases appear to be quite complex. Oxydo-reduction, acid-base and isomerization processes are intervening in the reaction path. Furthermore, a hydride ion has peculiar electronic structure which demands careful theoretical treatment. These may be some reasons that could have prevented theoretical studies on these enzymes in the past. For the same reasons they put exciting challenges to the theoretician presently.

4.2 Quantum studies on simple model systems.

There are several molecular orbital studies of simple model systems that might turn out to be useful to unravel electronic aspects of the catalytic mechanism of LADH. The electronic structure of lowest lying excited states of dihydronicotinamides and related molecules[40] have been studied with semiempirical molecular orbital configuration interaction method of Pariser, Parr and Pople. Portions of the conformational map of pyridine coenzymes and their constituents have been calculated with PCILO and STO-3G methods.[41] Zinc-ligands interactions have been calculated with rather reliable molecular orbital methods.[42] For the hydride-transfer reaction, MINDO/3 and STO-3G calculations on analogue reactions with cyclopropene, tropilidene and 1,4-dihydropyridine as hydride donors and cyclopropenium cation as acceptor have been carried out;[43a] The role of the $CONH_2$ group on the stereospecificity has also been examined.[43b] The electronic structure of the hydride transfer between LiH and cyclopropenium cation has been studied with STO-3G and 4-31G ab-initio procedures.[45]

Work on Zn-ligand interactions and the hydride transfer step are discussed in the following.

4.2.1 Zn-ligand interactions. Ab-initio quantum chemical studies on zinc liganded to imidazole, imidazolate anion, $NH_3$, $H_2S$, H5, $H_2O$ and $OH^-$ have been carried out by A. Pullman et al.(42) These ligands either appear in the catalytic zinc coordination sphere ($H_2O$, $OH^-$) or are models for the side chains of histidine and cysteine $(SH)^-$. In table 3 a selection of their results is presented. They concern equilibrium distances (R), proton affinities (PA) and charge transfer towards the zinc atom($\delta Q$).

Table 3

| Zn-ligand | $R_A$ | $R_{PL}$ | $PA_A$ | $PA_{FL}$ | $-\delta Q_A$ | $-\delta Q_{PL}$ |
|---|---|---|---|---|---|---|
| $OH^-$ | 1.70 | 1.90 | 82 | 380 | 0.509 | 0.271 |
| $Im^-$ | 1.82 | 1.87 | 168 | 364 | 0.658 | 0.460 |
| $SH^-$ | 2.09 | 2.33 | 85 | 354 | 0.557 | 0.320 |
| $(H_2O)OH^-$ | 1.72 | 1.91 | 110 | - | 0.429 | 0.173 |
| $(H_2O)_3OH^-$ | 1.80 | 1.98 | 156 | - | 0.282 | 0.119 |
| $(NH_3)_2ImOH^-$ | 1.80 | 2.1(Im) | 171 | 220(Im) | 0.301 | 0.134(Im) |

*Energy entries are in kcal/mol; distances in Å; charge transfer in units of electron charge. A=anion; PL=protonated ligand; FL=free ligand. The proton affinity quoted for imidazole free ligand is the experimental value.*

Some interesting trends can be observed in table 3. Zinc binding affects, in a significant manner, the proton affinities of the ligands. The changes are of qualitative and quantitative nature: The relative order to base strength for the free ligands is reshuffled by the cation - ligand interaction; There is a substantial lowering of the proton affinity itself. Thus, water and sulfhydric acid becomes stronger acids than imidazole, while the free ligands are all clustered around a proton affinity of 360 kcal/mol. The deprotonation of imidazole liganded to zinc becomes improbable when comparison is made with water or $H_2S$.

Changes in the coordination sphere bring about changes in the individual ligands. Anion proton affinity increases with ligands number, as well as ligand cation distance. The total charge transferred to zinc decreases with increasing ligand number. The nucleophilicity of bound hydroxyl may thus result from the nature, number and distances of the ligands entering in the coordination sphere.

Extensive ab-initio studies on the hydration spheres of zinc and transition metal cations have been reported by Rode et al.[46] Of particular interest is their work on the exchange processes between molecules belonging to the first and second solvation shells. For zinc cation a free activation energy for water exchange of 7.0 kcal/mol is reported. A thermodynamical analysis led these authors to conclude that model based on elimination of one solvent molecule with formation of a lower coordinated transition state was not satisfactory. Instead, a transition state with an increased coordination number, where the "descent" of a water molecule from the second shell towards the first one is accompanied by a simultaneous "ascent" of one molecule in the first shell towards the second one, was retained as a satisfactory model.

4.2.2. The hydride transfer step. The first problem to elucidate is the electronic structure of the hydride ion. For an organic chemist it is a proton with two electrons: $H^-$. From the quantum mechanical side a double zeta basis set is needed to describe it. Roothaan and Weiss[47] already in 196 have shown that $H^-$ is composed primarily of an electron weakly bound to the field of a polarized hydrogen atom, rather than having two simultaneous equivalent electrons, like a swollen $1s^2$ model of an helium atom. This theoretical fact puts severe constraints for selecting basis set: a minimal basis set cannot be used to prove or disprove the nature of an hydride transfer step.

Recently, MINDO/3 and STO-3G calculations on reactions with cyclopropene, tropilidene and 1,4-dihydropyridine as hydride donors and the cyclopropenium cation as acceptor have been reported. These reactions are analogues for the hydride-donation reaction of reduced nicotinamide adenine dinucleotide.[43a]

The intermolecular hydride transfer from cyclopropene to the cyclopropenium cation is best accomplished for a supermolecule of $C_s$ symmetry containing a linear C••••H•••C fragment. The reaction would consist of a concerted an gradual breaking of the C-H in cyclopropene, and the formation of a new C•••H bond in the cation fragment. At the transition state, the net atomic charge at the "hydride" is -0.006 ec (STO-3G) and -0.133 (MINDO/3). For all model reactions studied there is no evidence that at any particular instant a complete hydride ion is donated by one and accepted by the other reactant.

The preceding conclusion is based on minimal basis set calculations. STO-3G basis set permits the calculations of rather big molecules with reasonable computing times. In order to asses more quantitatively ability of this method to describe a typical hydride transfer we have carried out a series of calculations[45] on model compounds with 4-31G, 6-31G and STO-3G.

Lithium hydride has been taken as a test molecule. There are two reasons for this choice: first, it is a prototype of hydride donor source; second, there are available multiconfigurational selfconsisten-calculations, MSCF, which are representative of the state of the art in quantum chemistry.(48)

In table 4 the excess charges on the hydrogen atom at different interatomic distances calculated with the above mentioned methods are reported.

Table 4

| | $r_0$(Å) | $\Delta Q(r_0)$ | $\Delta Q(r+0.5)$ | $\Delta Q(r_0+0.8)$ |
|---|---|---|---|---|
| MSCF$^x$ | 1.5958 | -0.430 | -0.440 | -0.420 |
| 6-31G | 1.6369 | -0.264 | -0.263 | -0.267 |
| 4-31G | 1.6369 | -0.264 | -0.267 | -0.275 |
| STO-3G | 1.6369 | +0.015 | -0.012 | -0.043 |

*$r_0$ is the distance nearest to the minimum of the interatomic potential curves; $\Delta Q(r)$ is the excess Mulliken effective charge at distance r.*

*$^x$MSCF entries were kindly communicated to us by Dr Bo Jönsson.*

The results of table 4 show that STO-3G description of the charge density in the volume occupied by the hydrogen is unreliable. The results provided by the splitted valence electron shell method are in better agreement with the exact multiconfiguration results, although they are underestimated by about 40%.

To achieve comparisons between 4-31G and STO-3G basis sets a calculation of the hydride ion transfer between cyclopenyl cation and lithium hydride was carried out(45) Some results are summarized in table 5.

Table 5

| | 4-31G | | | STO3G | | |
|---|---|---|---|---|---|---|
| $R_{C-H}$ | $-E_{scf}$ (au) | $\theta_{opt}$ | $\Delta Q_H$ | $\Delta Q_H$ | $\theta_{opt}$ | $-E_{SCF}$ (au) |
| 1.9683 | 122.827970 | 77.84 | -0.395 | -0.074 | 76.62 | 121.502180 |
| .8683 | .832843 | 82.31 | -0.379 | -0.076 | 81.35 | .502822 |
| .7683 | .836835 | 86.49 | -0.352 | -0.077 | 81.85 | .503390 |
| .6683 | .840857 | 90.36 | -0.315 | -0.076 | 89.97 | .505092 |
| .5683 | .845516 | 93.89 | -0.271 | -0.073 | 93.70 | .508629 |
| .4683 | .851087 | 97.08 | -0.219 | -0.067 | 96.96 | .514107 |
| .2683 | .863456 | 102.39 | -0.145 | -0.047 | 102.18 | .527279 |

*The lithium-carbon distance is 3.5 Å; θ is the angle between the axis C-Li and the C-H bond at the vertex of cyclophenyl cation:*

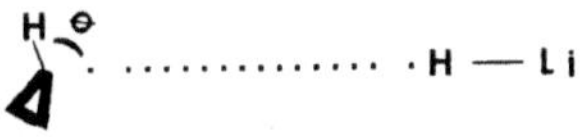


It can be seen from table 5 that the energy profile coming out from both basis set is rather similar: for the chossen path there is no energy barrier for the transfer of the hydrogen from lithium to cyclopropenyl cation. The absence of barrier is due to angular optimization carried out at each point of the potential energy curve. As far as the effective charge on the translocating hydrogen is concerned, the results differ. 4-31G description would rather correspond to the transfer of an hydride ion, while for STO3G there would be no hydride at all.

The point of controversy mentioned above has not only theoretical value. If an hydride ion is being actually transferred during the reaction, the presence of a water molecule may be catastrophic. The reason is simply given by the thermodynamically and kinetically rather fast reaction [49]

$$H^- + H_2O \longrightarrow H_2 + OH^-$$

Therefore, a simple rationale can be found to the fact that water is evacuated from the neighbourhood of the nicotinamide ring when the substrate is bound to the catalytic zinc.

It is thus desirable to carry out further theoretícal studies, with appropriate methods, to settle this important issue.

MINDO/3 and STO-3G calculations have also been used by the same authors to study the role of the $CONH_2$ group for the stereospecificity of hydride transfer to the nicotinamide ring.[43b] The authors conclude that it seems possible to relate the stereospecificity simply to an out-of-plane orientation of the $CONH_2$ group in the transition state. It is surprising, to say the least, that these authors do not even mention the fact that X-ray crystallographic studies of alcoholdehydrogenase,[34] lactate dehydrogenase[35] and glyceraldehyde-3-phosphate dehydrogenase[36] have given a perfectly simple and reasonable explanation for this stereospecificity in geometrical terns.[50] In the first two enzymes which are A-stereospecific, nicotinamide rings are bound to the enzymes in such a way that the A-side of the ring faces the substrate, whereas the B-side faces the protein wall. In glyceraldehyde-3-phosphate dehydrogenase, which is B-stereospecific the opposite is true.

### 4.3. Quantum/statistical studies on LADH.

A variety of molecular motions are involved in the catalytic mechanism of LADH. The description of proton translocation after coenzyme binding, hydride transfer and electronic structure of zinc during the reaction requires a quantum chemical framework. Domain movements, substrate and nicotinamide librations are subjects for molecular dynamics. Water structure inside the protein can be simulated with statistical mechanical Monte Carlo simulation procedures.

In this section quantum chemical studies on the proton relay structure and Monte Carlo simulations of water in the substrate channel of LADH are described.

4.3.1 Proton relay system. An H-bonded triad of imidazole-methanol-water has been taken as a cage system to represent the proton relay structure. The protein core includes the catalytic zinc cation with fractional charge +0.5, a positive charge at the position of the (ring) nitrogen atom of the nicotinamide moiety, and all atoms of the corresponding subunit. The intermolecular distance used are those found by X-ray crystallography.

The proton relay process is basically associated with the passage from canonical H-bonded structure, $H_2O\cdots CH_3OH\cdots Im$, to the ion-pair form, $OH^-\cdots CH_3OH\cdots ImH^+$. Proton potential curves were calculated for a variety of screened Zn cation; the results with $Zn^{+0.5}$ are illustrative of the trends obtained in all the calculations.[7,29]

The calculations are carried out in steps:

i) supermolecule in vacuo, including the +0.5 charge at zinc-atom place
ii) supermolecule plus the protein permanent field
iii) supermolecule in protein core without the +1 charge of the coenzyme
iV) inclusion of the coenzyme via a positive charge.

The proton potential curves for steps i) to iV) are drawn in fig. 2.

Fig. 2.

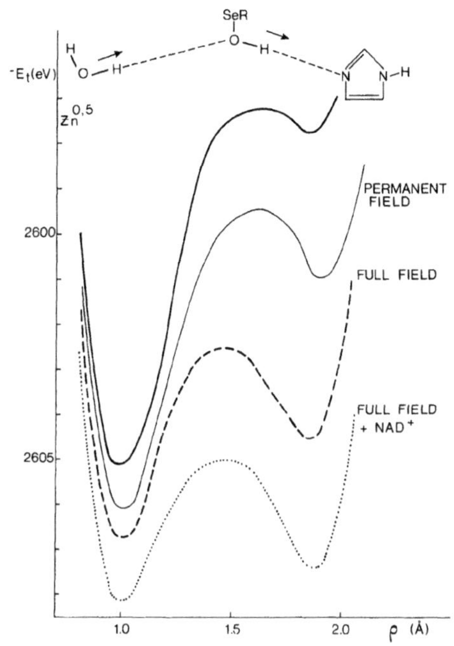


*Total energy as a function of the simultaneous displacement coordinates for the protons in the bridge. Upper curve describes the system in vacuo.*

The different calculational steps represented in fig. 2 serve to illustrate the sensitiveness of the proton relay structure to the protein potentials. One can appreciate that both the permanent and polarization protein potentials are essential to stabilize the ion-pair structure. The triggering effect of $NAD^+$ is illustrated by the lower curve of this figure. These calculations show that the protein core potential would favour the evacuation of a positive charge from the interior of the protein towards its surface. This is an important piece of information to understand the catalytic mechanism of LADH, since this mechanism requires on the one hand to screen the reactants from the solvent and on the other to separate a proton from the same reactants.

4.3.2. Water structure simulations. An important piece of information to describe the molecular mechanism of LADH is the knowledge of the water structure in the substrate hydrophobic barrel after coenzyme binding. It is important to know if there are water molecules at the active site loosely bound to the coenzyme nicotinamide moiety and or in the second coordination shell of zinc. At the present resolution the X-ray studies are somewhat inconclusive on that point. To obtain additional information on this point statistical mechanical Monte Carlo simulations of water structure in this region were made.[51] We briefly describe the model and some important results.

The model system is built up from the enzyme and coenzyme atoms which are found inside a cylinder of 5 Å radius with an axis along the outward direction of the barrel. Coordinates from the triclinic LADH-NADH-DMSO (dimethyl sulphoxide)(52) were taken as input data. 25 water molecules, with an ice-structure, were set up in a volume outside the entrance to the barrel. The temperature was kept at 300°K.

The protein structure corresponds to the holo form. The isomerization from apo- to holo- and subsequent DMSO binding narrows the entrance to the active site cleft. We wanted to know whether the cleft has been fully isolated from the solvent, or if the channel leaves enough space for water molecules to enter or leave.

Thus the following model situation was build up: DMSO, which is bound to zinc, was removed; the oxidized form of the coenzyme was simulated with a +1 unit charge at the nitrogen position in the nicotinamide ring; a-1 unit charge was set up at the sulfur atoms of Cys 46 and 174; a +2 charge unit was given to the zinc; Clementi potential for water-water interaction(53) was used; protein-water interactions were represented by a repulsive hard core (van der Waals radii) and charge-dipole interactions. The effective charges used to build up the permanent protein potential were employed for this purpose. This elementary protein-water interactive potential permits to accelerate the calculations.

In practice, thermal equilibration is attained after $10^6$ moves; another $10^6$ configurations are thence generated and stored to carry out the statistical analysis.

We summarize four important results of these calculations.

1) The channel is not fully closed to water molecules;

2) One water molecule moves from about 15 Å away from the zinc at the beginning of the simulation into the zinc coordination sphere, to 2.4 Å,occupy the position of the zinc bound water in the apoenzyme structure.

3) No additional water molecules enter the coordination sphere of zinc confirming mechanistic models based on tetracoordinate zinc.

4) The radial distribution functions, which are measured from the catalytic zinc position, as well as the averaged coordinates of water sample show a network of water molecules. These are at H-bonding distances from each other starting at the water bound to zinc and extending into the bulk through the hydrophobic channel.

5) No water molecule is in close contact to the nicotinamide ring.

## 5. GENERAL DISCUSSION

An overview of the inhomogeneous selfconsistent reaction field theory of protein core effects has been presented, and applications to the catalytic mechanisms of serine proteases and alcohol dehydrogenase have been described.

The theory provides a framework where methods and informations from protein physical chemistry, statistical mechanics and quantum chemistry can be integrated in a natural way. The calculations of chemical reactivity indices,potential surfaces of reactive coordinate paths, and electronic properties of reactant and probe systems in the presence of the protein environment can be made. Differential effects produced by the protein (solvent) can be assesed by calculating the quantum subsystem as if it were in vacuo.

The ISCRF theory has been implemented with approximate molecular orbital (MO) methods. In the applications adequate corrections for proton affinity errors have been taken into account. In general, such approximate MO schemes describe H-bonding and proton potential curves. The trends found when the protein effects are incorporated are physically and biochemically meaningful. In some cases they have been confirmed by ab-initio calculations where protein effects are also incorporated, albeit differently than ours. Furthermore, they agree with present experimental knowledge.

Statistical mechanical Monte Carlo simulations of water structure in LADH have shown the potentiality of this method. Important information concerning the coordination sphere of the catalytic zinc and distribution of water in the substrate hydrophobic barrel has been obtained.

The description of the hydride transfer step in LADH has been more difficult. The conspicious electronic structure of the isolated hydride ion prevents using standard minimal basis set for quantum calculations. We have shown with splitted valence shell orbital methods that for some model compounds actual transfer of an hydride ion is quite probable. Further studies are needed to make more definite statements on this issue. In particular, protein core effects have to be taken into account. They may play an important role in the actual enzyme reaction.

At the present state of development the ISCRF theory of protein core effects is becoming a useful tool to provide supplementary information at the electronic level on enzyme catalytic properties.

## REFERENCES

1. B. Pullman (ed), "Catalysis in Chemistry and Biochemistry. Theory and Experiment" (Reidel, Dordrecht, 1979).
2. H. Weinstein and J.P. Green (eds), "Quantum Chemistry in Biomedical Sciences" (Published by the New York Academy of Sciences 1981).

3. M.R. Pincus and H.A. Scheraga, Acc. Chem. Res. 14, 299 (1981).
4. G. Naray-Szabó and T. Bleha, in "Progress in Theoretical Organic Chemistry" (Ed. by I.G. Csizmadia, Elsevier, 1982). In press.
5. A. Warshel, Biochemistry 20, 3167 (1981).
6. A. Warshel, Acc. Chem. Res. (1981).
7. O. Tapia and G. Johannin, J. Chem. Phys., 75, 3624 (1981)
8. O. Tapia in "Molecular Interactions", 3, ch. 2 (ed. by Daudel, R., Pullman, A., Salem, L. and Veillard, A.) (Reidel, Dordrecht, 1980). 2, pp 25-72.
9a. E. Clementi and G. Corongin, Ann. N.Y. Acad. Sci., 367, 83 (1981).
9b. D.L. Beveridge et al., Ann. N.Y. Acad. Sci., 367, 108 (1981).
10a. P.J. Rossky and M. Karplus, J.Am. Chem. Soc., 101, 1913 (1979).
10b. M. Karplus, Ann. N.Y. Acad. Sci., 367, 407 (1981).
11. J.A. Pople and D.L. Beveridge, "Approximate Molecular Orbital Theory" (McGraw Hill, New York, 1970).
12. R. Pethig, "Dielectric and Electronic Properties of Biological Materials". (Wiley, Chichester, 1979).
13a. D.M. Blow, Acc. Chem. Res., 9, 145 (1976)
b. J. Kraut, Ann. Rev. Biochem. 46, 331 (1977)
14a. D.A. Matthews, R.A. Alden, J.J. Birktoft, S.T. Freer and J. Kraut, J. Biol. Chem., 252, 8875 (1977).
14b. M.N. James, Can. J. Biochemistry, 58, 251 (1970).
15. M. Komiyama and M.L. Bender, Proc. Natl. Acad. Sci. USA, 76, 557 (1979).
16. W.W. Bachovchin and J.D. Roberts, J.Am. Chem. Soc., 100. 8041 (1978).
17. M.N.G. James, Can. J. Biochem., 58, 251 (1980).
18. S.E. Hamilton and B. Zerner, J.Am.Chem.Soc., 103, 1827 (1981).
19. A.A. Kossiakof and S.A. Spencer, Biochemistry 20, 6462 (1981).
20. S.A. Bizzozero and H. Dutler, Bioorg. Chem. 10, 46 (1981).
21. H. Umeyama and S. Nakagawa, Chem. Pharm. Bull. 27, 1524 (1979).
22. S. Nakagawa, H. Umeyama and T. Kudo, Chem. Pharm. Bull., 28, 1342 (1980).
23. H. Umeyama, S. Nakagawa and T. Kudo, J. Mol. Biol., 150, 409 (1981).
24. P. Banacky and B. Linder, Biophys. Chem., 13, 223 (1981).
25. D.M. Hayes and P.A. Kollman, see page 77 in ref. 1.
26. L.C. Allen, Ann. N.Y. Acad. Sci., 367, 383 (1981).
27. D.F. deTar, J. Am. Chem. Soc., 103, 107 (1981)
28. O. Tapia, A.-M. Armbruster, F. Stamato, E. Largo, G. Johannin and C.-I. Brändén (1982) To be published.
29. O. Tapia, C. Lamborelle and G. Johannin, Chem. Phys. Lett., 72, 334 (1980).
30. L. Sawyer, D.M. Skotton, J.W. Campbell, P.L. Wendel, H. Muirhead, H.C. Watson, R. Diamond and R.C. Cadner, J. Mol. Biol., 118, 137 (1978).
31. J.J. Birktoft and D.M. Blow, J. Mol. Biol., 68, 187 (1972).
32. J.A. Pople in "Electronic Structure Theory" vol. 4 (Ed. by H.F. Schaefer III, Plenum Press, New York, 1977).
33. L.C. Allen, Ann. N.Y. Acad. Sci., 367, 383 (1981).
34. H. Eklund, B. Nordström, E. Zeppezauer, G. Söderlund, I. Ohlsson, T. Boiwe, B.-O. Söderberg, O. Tapia, C.-I. Brändén and Å. Åkeson, J. Mol. Biol., 102, 27 (1976).

35. W. Eventoff, M.G. Rossman, S.S. Taylor, H.-J. Torff, H. Meyer, W. Keil and H.-H. Kiltz, Proc. Natl. Acad. Sci. (USA) 74, 2677 (1977).
36. M. Buehner, G.C. Ford, D. Moras, K.W. Olsen and M.G. Rossman, J. Mol. Biol. 82, 563 (1974).
37. C.-I. Brändén and H. Eklund, in "Dehydrogenases Requiring Nicotinamide Coenzyme" (Ed. J. Jeffery, Birkhäuser Verlag, Basel, 1980).
38. A. Fersht, "Enzyme Structure and Mechanism" (Freeman, Reading, 1977).
39. H. Dutler, in "Pyridine Nucleotide Dependent Dehydrogenases" (Ed. H. Sund, Walter de Gruyter, Berlin, 1977) pp. 339.
40. E.M. Evleth, J. Am. Chem. Soc., 89, 6445 (1967).
41. D. Perahia, B. Pullman and A. Saran, in "Structure and Conformation of Nucleic Acids and Protein-Nucleic Acid Interactions", (Ed. by M. Sundaralingam and T.S. Rao, University Park Press, Baltimore, 1975) pp. 685.
42a. D. Demoulin, A. Pullman and B. Sarkar, J.Am.Chem.Soc., 99, 8498 (1977).
42b. D. Demoulin and A. Pullman, Theoret. Chim. Acta 49, 161 (1978).
42c. A. Pullman and D. Demoulin, Int. J. Quantum Chem., 16, (1979).
42d. V. Kothekar, A. Pullman and D. Demoulin, Int. J. Quantum Chem. 14, , (1978)
43a. M.C.A. Donkersloot and H.M. Buck, J. Am. Chem. Soc. 103, 6549 (1981).
43b. M.C.A. Donkersloot and H.M. Buck, J. Am. Chem. Soc., 103, 6554 (1981).
44. J. Kvassman, A. Larsson and G. Pettersson, Eur. J. Biochem., 114, 555 (1981)
45. O. Tapia, to be published
46. B.M.Rode, G.J. Reibnegger and S. Fujiwara, JCS Faraday II, 76, 1268 (1980).
47a. C.C.J. Roothaan and A.W. Weiss, Rev. Mod. Phys., 32, 194 (1960).
b. For ab-initio studies on reactions of the hydride ion with small molecules see also: L. Radom, in "Electronic Structure Theory", (Ed. by H. F. Schafer III, Plenum, 1977) chp. 8, in particular pp. 342-352, and references cited therein.
48. B. Jönsson, B.O. Roos, and P. Taylor, J.Chem.Phys., 74, 4566 (1981).
49. T. Moeller, "Inorganic Chemistry" (Wiley, 1953) pp. 408.
50. M.E. Rossman, A. Liljas, C.-I. Brändén and L. J. Banaszak, in "The Enzymes", (Ed. by P.D. Boyer) Academic Press, New York, 1975).
51. B. Jönsson, O. Tapia and C.-I. Brändén, to be published.
52. H. Eklund and C.-I. Brändén, J. Biol. Chem., 245, 3458 (1979)
53a. O. Matsuoka, M. Yoshimine and E. Clementi, J.Chem.Phys., 64, 1351 (1976).
53b. E. Clementi, F. Cavallone and R. Scordamaglia, J.Am.Chem.Soc., 99, 5531 (1977).
53c. S. Romano and C. Clementi, Gaz.Chim.Italiana, 108, 319 (1978).

# APPLICATIONS OF QUANTUM CHEMISTRY TO PHARMACOLOGY

W. Graham Richards
Physical Chemistry Laboratory, Oxford University, England

ABSTRACT

Quantum mechanical calculations may focus attention either on the energy of a molecule or the electronic wave function resulting from the computation. Energies have been used to determine tautomer ratios and equilibrium conformations, but are of most use when answering questions as to the possible shapes which a molecule may adopt if distorted on binding. The wave function leads to molecular descriptions in electronic terms. Only with the introduction of computer graphics, particularly involving colour, have the full potentials begun to be realised.

## 1. INTRODUCTION

Until very recently practitioners of quantum pharmacology were teased by sceptics who asked just how many drugs have been discovered by these methods. The answer is still that none has emerged solely as a result of theory. However the computational chemist has contributed to the extent that in a number of instances they have been included on patents alongside synthetic organic chemists. Virtually all the major pharmaceutical houses now employ experts doing theoretical work so that it is no longer necessary to be on the defensive about the subject.

All quantum mechanical calculations start from the Schrödinger equation,

$$H\Psi = E\Psi$$

yielding an energy and a wave function for a given input of nuclear coordinates. Earlier work concentrated on energy calculations while currently the wave function receives more attention. In both types of work the display of the calculated information in a form which can be assimilated by a medicinal chemist is a serious problem, and in both areas computer graphics have been profoundly influential.



*R. Daudel, A. Pullman, L. Salem, and A. Viellard (eds.), Quantum Theory of Chemical Reactions, Volume III, 125–134.*

## 2. CALCULATIONS BASED ON ENERGY

Accurate molecular geometries are best found from X-ray crystallography but there are now available quantum mechanical methods which can optimize geometries very successfully. An example is the HONDO program (Dupuis et al. 1981).

Tautomer ratios are sometimes susceptible to nmr studies but if only one form is observed then resort may be made to calculation. However in such problems, extreme care must be made to optimize the geometries of each form or spurious results may emerge. For example we have found that for tautomers of imidazole there are drastic changes in the bond lengths and angles of the ring when a proton is transferred from one ring nitrogen to the other (Richards, et al. 1979).

In the area of conformation, calculations can add significantly to experiment. If for the moment we restrict our attention to instances where conformation is defined by two torsion angles (a limitation which still allows consideration of acetylcholine, catecholamines, histamine, and GABA) then a two-dimensional energy map gives all the necessary conformational data. On such a surface crystallography reveals a single point; nmr highlights small sections of the surface populated sufficiently to yield a signal, while theory may describe the entire surface.

Because of approximations such as neglect of solvent it is frequently best to use conformational potential surfaces in a comparative study of physically and chemically similar molecules which elicit similar biological responses of varying strengths. In order to make comparisons and to permit answers to the question, "Is there a specific range of conformation required for binding to a particular receptor?", it is helpful to convert energy surfaces into population maps. These may then be used in Venn diagrams as in Figure 1.

We first used this technique to define an essential conformation for activity at the histamine H1 receptor (Farnell et al. 1975). Perhaps more interesting is the case of H2 activity. Histamine itself is active at both receptors, while 4-methyl histamine is barely active at H1 and about half as active as the natural compound at H2. The $\alpha$-methyl substituted histamine is scarcely active at either site. However, the dimethylated compound, $\alpha$-methyl 4-methyl histamine has greatly increased H2 activity with respect to $\alpha$-methyl. This is a rare example of an increase of activity on methylation; or steric assistance rather than steric hindrance. Histamine is a particularly awkward molecule from the point of view of calculations. Isolated molecule calculations indicate strong intramolecular hydrogen bonding which would not be present in aqueous solution. Thus it is not possible to exploit the experimental facts fully, but the Venn diagram approach does suggest that it may be a prerequisite that the molecule can achieve or at least pass through a planar conformation in order to bind successfully to the receptor.

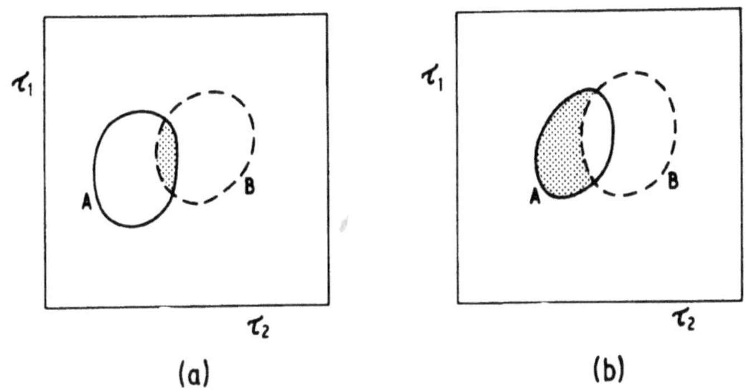


Figure 1. Venn diagrams showing conformations necessary but not sufficient for activity (shaded)
a) Molecules A and B both active
b) Molecule A active but B inactive

Similar studies can be made for dopamine (Figure 2) and for GABA (Figure 3), but in these instances the major lack is on the experimental side where data such as those for histamine are not available. Nonetheless, tentative conclusions can be drawn.

If the molecules of interest have several torsion angles then populations may no longer be displayed as simple maps. However it is still possible to use the Venn diagram approach but now by comparing each molecule of a series with a template whose conformation is thought to be known.

An excellent example is furnished by the work of Dr K. Heritage of ICI Plant Protection Division on synthetic pyrethroids. The basic skeleton is shown in Figure 4.

The activity of some related molecules with aromatic substituents can be rationalised if we posulate that the molecules adopt a form close to the minimum conformation of the template molecule. For each molecule in the series it is then possible to produce a Venn diagram with three areas: one represents conformational flexibility; one the closeness of two aryl rings and the third the coplanarity of these rings. Active molecules should have the areas overlapping while less

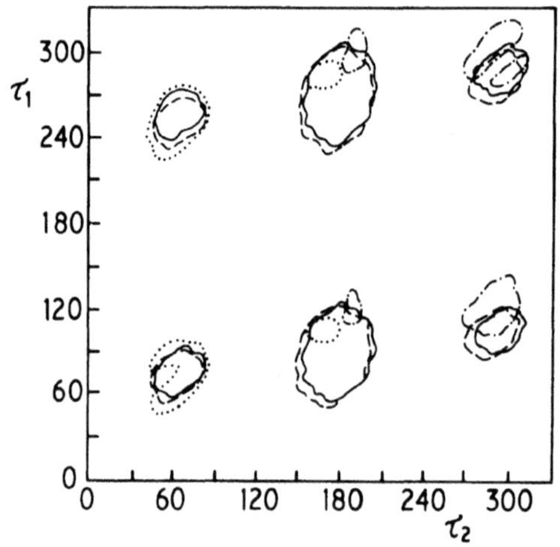


Figure 2. Population maps for dopamine and substituted dopamines. Dopamine and N-methyl dopamine are equipotent but α- and β-methyl compounds are less active. The only shape which appears to discriminate is the antiperiplanar conformation ($\tau_1 = 90^\circ$ or $270^\circ$; $\tau_2 = 180^\circ$)

potent species cannot fulfil all three criteria simultaneously.

The conformational calculations for molecules with many degrees of freedom may be performed either by quantum mechanical methods or empirical molecular mechanics techniques. In either case, (and possibly a combination is the best approach), the use of computer graphics to display the molecules is essential. It is also advantageous to be able to rotate the molecules in real time on the screen. If in addition it is helpful to superimpose molecules to test for similarity in shape then colour adds very significantly to the clarity of both the picture and its interpretation.

## 3. CALCULATIONS BASED ON THE WAVE FUNCTION

Although totally empirical methods can be used for conformational problems which depend solely on the disposition of nuclei, only quantum mechanical calculations reveal electronic descriptions. The obvious importance of this aspect lies in the fact that the molecular events are controlled by electrons and the nuclear framework merely provides the skeleton which the electron density fleshes out.

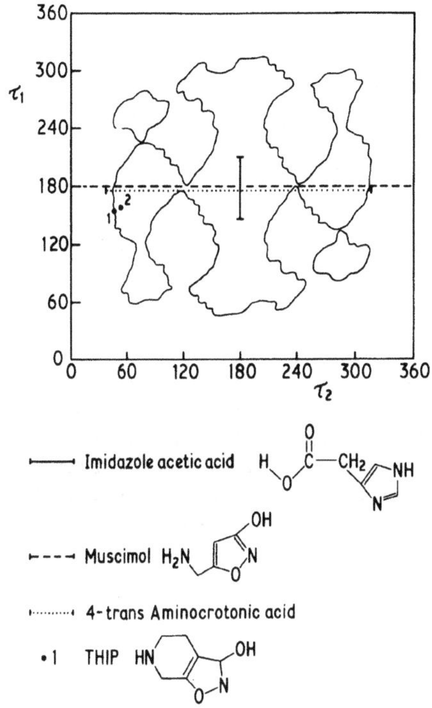


Figure 3. Population map for GABA with available conformations for some agonists.

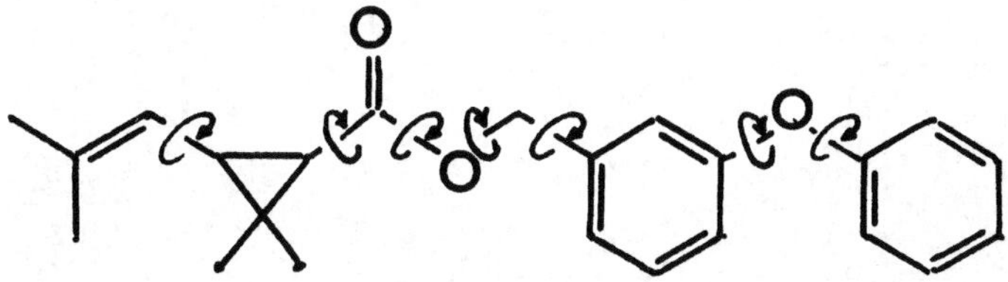

Figure 4. A synthetic pyrethroid showing the seven torsion angles.

Since electron density is given by the value of $(\Psi^*\Psi)$ at any point the subtlties of the wave function are not as important for charge distribution as they are for expectation values of other operators. This means that even rather crude or semi-empirical wave functions can be used to calculate a charge density. The major problem is not the computation which is almost trivial, but the display Obviously charge density can be shown as a contour diagram such as is to be found in elementary text books. However, for the non-symmetrical molecules founo in pharmacology there is frequently no obvious plane in the molecule in which to draw the density contours. The advent of computer graphics systems has permitted the complexity of visualisation to be overcome, particularly if colour and 'depth cueing' are employed.

The nuclear framework is displayed as in a Dreiding model and the electronic 'shape' presented as a net covering the exterior of a molecule at a prescribed electron density contour. The choice of the contour is arbitrary but a logical choice may be made from the consideration of separate quantum mechanical computations of molecular interaction energies. The interaction of a helium atom or a water molecule has a potential which becomes steep or repulsive at about 0.001 atomic units of charge for the isolated molecule. Figure 5 illustrates pyrrole in these terms

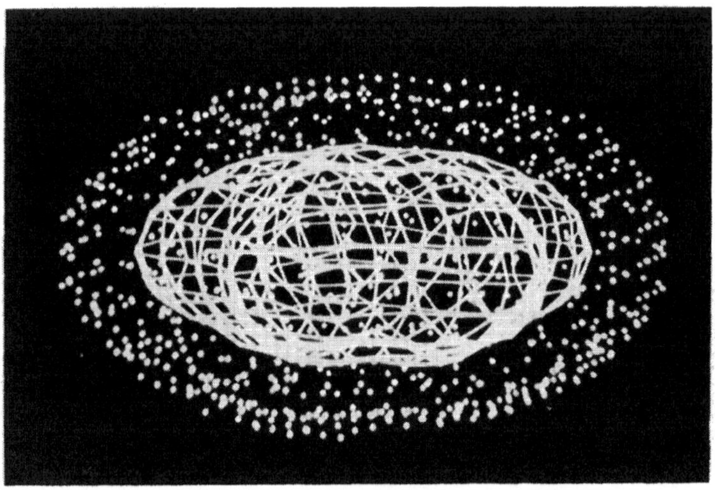

Figure 5. Charge density contours surrounding pyrrole. Solid line at 0.01 atomic units of charge; dots at 0.001 atomic units.

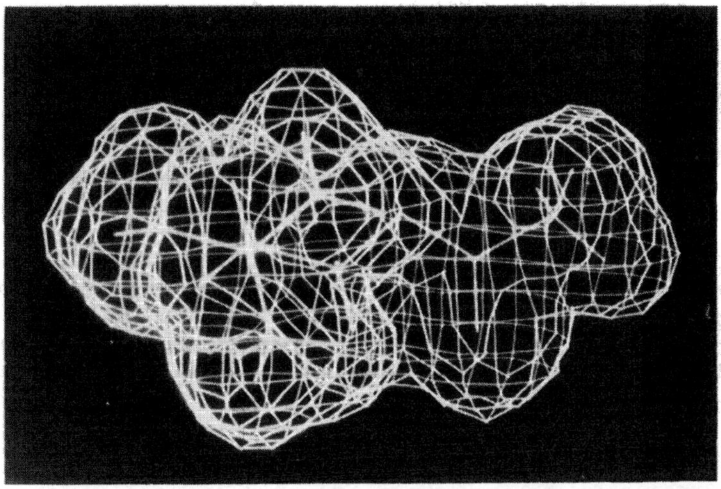

Figure 6. a) Acetylcholine

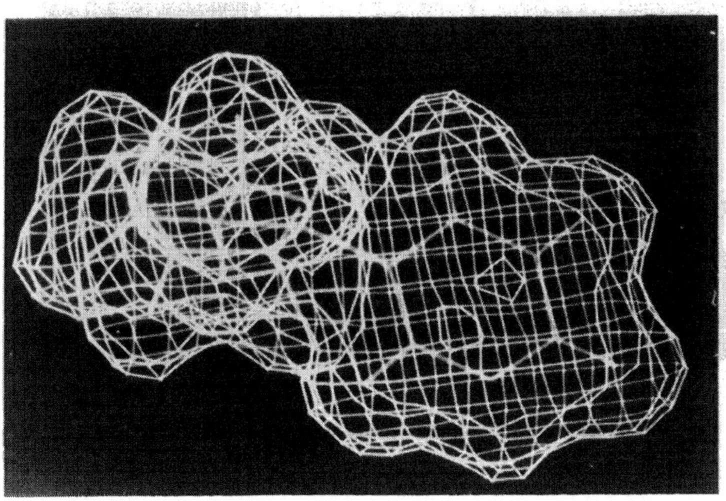

Figure 6. b) Nictotine

A net drawn over a molecule can then indicate similarities in shape which are not obvious from space-filling models. Figure 6 illustrates the similarity between actylcholine and nicotine; a similarity which has been extended to commercial compounds.

As an alternative to charge density the equivalent but more revealing electrostatic molecular potential may be used. Two-dimensional maps of the potential generated by active molecules have been used by Weinstein (1979) in particular to indicate essential features of electronic structure required for pharmacological acitivity. Once again this work can be made more palatable to the working medicinal chemist of graphics facilities are used with positive and negative regions differentiated by colour. Figure 7 is a rather inadequate reproduction in black and white of a much more useful figure originally in colour.

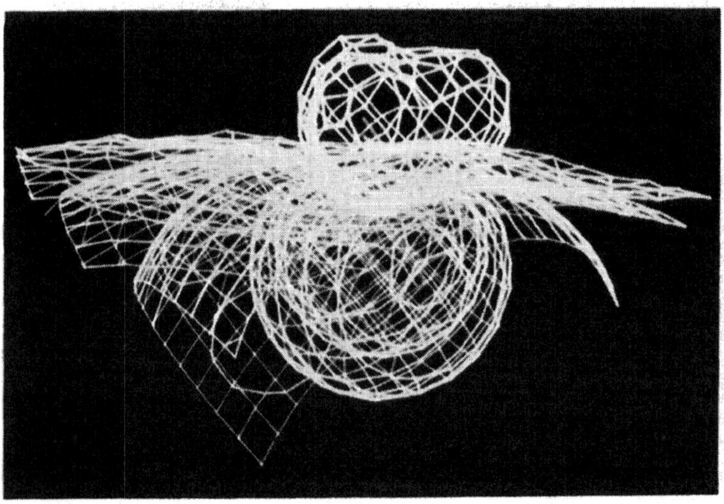

Figure 7. Electrostatic molecular potential of ammonia

There are two objections to these uses of the wave function. In the first place they do not produce numbers suitable for use in a regression analysis and secondly they take no account of just how a molecule might be modified by the atoms making up the receptor binding site.

The first problem may be overcome to some extent by throwing away some of the charge density detail and integrating the charge within a sphere of a defined radius which may be placed on a nucleus, in a bond, or at any suitable point in the molecular space (Dean and Richards,

1975; Wallis 1981). In a series of recent papers (see Elliott et al. 1982) we have shown that chemical reactivity is dependent on the total charge for 'hard' reagents and on the sphere charge in specific orbitals for 'soft' reagents. If molecules are donating charge then details of densities in the highest occupied molecular orbital produce quantitative correlations and similarly for acceptors the lowest unoccupied molecular orbital cominates. In this way it is possible to produce a form of quantitative frontier orbital theory. The published applications have concerned correlations with organic chemical reactivity, but private applications have revealed similar correlations for commercially interesting biological action.

Although in general there are insufficient data available to include details of receptor binding sites in theoretical calculations, the same is not true for enzymes where in a number of important instances X-ray crystallography and nmr have yielded atomic detail. Obvious pharmaceutical targets in the category include for example dihydrofolate reductase. To include all the atoms of a binding site in a full ab initio quantum mechanical calculation is beyond the limits of contemporary computing power. However, a technique deriving from the work of Kollman (1976) permits all the atoms of the binding site to be included if each is replaced by point partial charge. In effect, into the Hamiltonian for the problem we put in the known positions some dummy atoms which have non-integral nuclear charges and no electrons.

These enzyme calculations again almost demand the use of computer graphics (Langridge 1981) if they are to be exploited fully, especially in coping with relative orientations of substrate and site. In as yet unpublished work Sandra Robins in our laboratory has shown that the relative binding energies of blockers of chymotrypsin can be reproduced quantitatively to an accuracy of about 1 kcal.$mol^{-1}$.

## 4. CONCLUSIONS

The current trend in quantum pharmacology is to exploit computer graphic displays both in conformational energy calculations and in presenting electronic information. These techniques are at last putting theoretical methods into the hands of medicinal chemists who are not full-time computational chemists. The outcome has been a significant input into the search for new compounds; successful to the point of the patenting.

REFERENCES

Dean, S.M. and Richards, W.G.: 1975, Nature, 258 p133.

Dupuis, M., Rys, J. and King, H.F.: 1981, QCPE 13 p403.

Elliott, R.J., Sackwild, V. and Richards, W.G.: 1982, J. mol. Struct. THEOCHEM., 86 pp 301-314.

Farnell, L., Richards, W.G. and Ganellin, C.R.: 1975, J. med. Chem. 18 pp 662-5.

Kollmann, P.: 1976, J. Am.chem.Soc. 98 p.3335.

Langridge, R,Ferrin, P.E., Kuntz, I.D. and Connolly, M.L.: 1981 Science, 211 p 661.

Richards, W.G., Wallis, J. and Ganellin, C.R.: 1979, Eur.J.med.Chem. 14 p9.

Weinstein, H. in Computer assisted drug design (eds. Olson, E.C. and Christoffersen, R.E.) American Chem.Soc., Washington (1979).

# ON THE PHARMACOPHORE AND MODE OF ACTION OF SOME SCHISTOSOMICIDAL AGENTS. CONFORMATIONAL ASPECT.

Federico Peradejordi[+] and Elson Longo da Silva[++]
Centre de Mécanique Ondulatoire Appliquée (C.N.R.S.)
23 rue du Maroc, 75019 Paris

## I. Background

Schistosomiasis[(1)], estimated to affect more than 200 million people in the tropics[(2)], constitutes the most serious endemic parasitic infection in man.

The severe host toxicity of the trivalent organic derivatives of antimony, the oldest chemotherapeutic agents used in the treatment of schistosomiasis[(3)], led to the search for less toxic antischistosomal drugs. The discovery of the antischistosome activity of Lucanthone[(4)] (I) stimulated a great deal of structural modifications of the parent nucleus in an attemp to develop better agents[(5)], and several additional series of compounds representing or resembling portions of the active parent structure where synthesized[(6,7)]. Figure 1 shows the basic molecular nucleus for the most important compound series.

Although at first glance the structural resemblence of these series is not great, all the compounds under consideration do have certains common features. The results of extensive metabolic studies[(1,7)] on Lucanthone and its 4-methylated related congeners (II), (III) and (IV), in mammalian organisms and in vitro systems, indicate that they are hydroxylated in vivo and that the corresponding 4-hydroxymethyl metabolites (V), (VI), (VII), and (VIII) are responsible for their observed schistosomicidal activity. On the basis of structure-activity studies[(1,7,8)], it is generally conceded that the essential molecular structural features responsible for the schistosomicial efficacy in these compounds include an alkylaminoethylamino side-chain para on an aromatic ring to the hydroxymethyl group, which in turn must be flanked by a more electronegative substituent. The ring system may be modified to a certain extent without loss of efficacy. The introduction of additional substituents into the molecules sometimes causes changes in the

+ To whom inquiries should be addressed
++ Present address : Departamento de Química, Centro de Ciências e Technologia, Universidade Federal de São Carlos, Brasil.



*R. Daudel, A. Pullman, L. Salem, and A. Viellard (eds.), Quantum Theory of Chemical Reactions, Volume III, 135–149.*

activity. The basic side-chain may be varied over a limited range with little loss in biological activity.

In view of these results it is thought that all these series of compounds fall into the same therapeutical class and all members are probably acting by a similar mechanism[1]. Nevertheless the elucidation of the mode of action by which these drugs exert their lethal effects on schistosomes is still underway. Experimental results indicates[1,2,7,11,12] that the antischistosomal action of Hycanthone (V) is brought about by mechanisms which are different from those responsible for some effects of this drug on the host produced by interaction with base residues in DNA[13,14]. At present, an action by blocking Acethylcholine (IX) binding sites of schistosomes seems to constitute the most plausible

(IX)

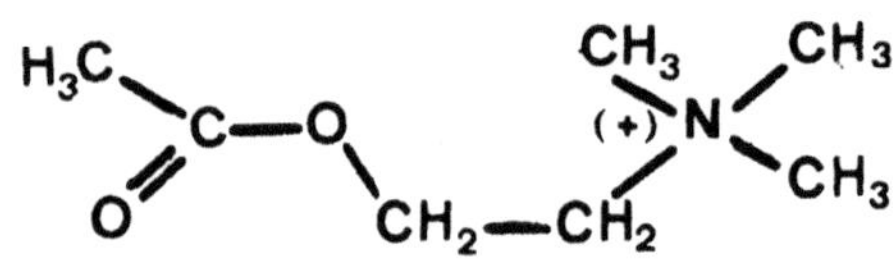


hypothesis for the mechanism of action of these compounds[9,10]. Acethylcholine (Ach) has been recognized[15] as a neurotransmitter controlling movements of the schistosomes oral sucker, the acetabulum (the muscular organ with which the worm attaches itself to the internal wall of a mesenteric vein) and the gut. All these substances are structurally similar to ACh with the ethylenediamine chain $-NR-CH_2-CH_2-\overset{+}{N}HR-$ equivalent, when protonated at the terminal nitrogen[14], to the choline fragment $-O-CH_2-CH_2-\overset{+}{N}Me_3$ and the planar acetoxy group replaced by a benzene ring. On the basis of experimental results concerning the effect of Hycanthone on schistosomes motor activity[16] and on fluorescence labelling of ACh binding sites[17], recent repports have suggested that Hycanthone is an inhibitor of schistosomes ACh receptors. This inhibition could lead to functional disturbances in the worms reflected i) in the reversible loss of attachement of schistosomes to the internal wall of the mesenteric veins and their shift towards the liver carried by the venous blood flow ; ii) in the observed slow, delayed death of parasites probably due to starvation[1,10,18,19] ensuing the functional impairment of the worms digestive apparatus. ACh receptors in schistosomes have some similarities but are apparently not identical with any of the pharmacologically well defined receptors of ACh in mammalians[15-17], these findings conforming with the selective activity of the drugs in schistosomes but not mammals.

Data summarized above support the proposed mechanism of action that chemotherapeutic effects of these drugs may presumably be accounted for by an impairment of schistosomes muscular activity following the blockade of their ACh binding sites. If this hypothesis proved to be correct, it may provide the rationale for the development of chemotherapeutic agents which will have the desired lethal effect on schistosomes, but which entail minimal or no human toxicity risks.

| Compound | Common preferred geometries | Interatomic distances (Å) | | | | Net charges[d] | | | | |
|---|---|---|---|---|---|---|---|---|---|---|
| | | $N_9$--$\overset{+}{N}_{12}$ | ring center--$\overset{+}{N}12$ | Cl(S)--$\overset{+}{N}$ | Cl(S)--O | $\overset{+}{N}$ | $(\overset{+}{N})^e$ | $N_9$ | Cl(S) | O |
| N-phenylpiperazinium | {0°, 180°[a],53,7°[a]} | 2.98 | 5.67 | 7.17 | 2.28 | 0.200 | 0.972 | -0.348 | -0.361 | -0.553 |
| Aminomethyltetrahydro-quinoline | {0°[a],208°[a], 60° } | 2.99 | 5.02 | 7.36 | 2.28 | 0.138 | 0.957 | -0.468 | -0.366 | -0.553 |
| Aminoethylaminoben-zene | {0°, 180° , 60° } | 2.88 | 5.56 | 7.64 | 2.28 | 0.129 | 0.970 | -0.450 | -0.361 | -0.553 |
| Hycanthone | {0°[a], 197.8°,76.5°}[b] | 3.03 | 5.60 | 8.10 | 1.79 | 0.217 | 0.983 | -0.459 | -0.405 | -0.343 |
| | | O--$\overset{+}{N}$ | | | | | | | | |
| ACh (muscarinic) | {0°,∿180° ,∿85° }[c] | 3.26 | | | | | | | | |
| ACh (nicotinic) | {0°,∿180° ,∿75° }[c] | 3.07 | | | | | | | | |

Table 1. Summary of conformational and net charge data.

[a]Constrained coordinates
[b]Ref. 23
[c]Refs. 26, 28
[d]CNDO/2 calculated values for solvated molecules
[e]Cationic head charge : sum of net charges on $\overset{+}{N}_{12}$ and its four attached neighbouring atoms

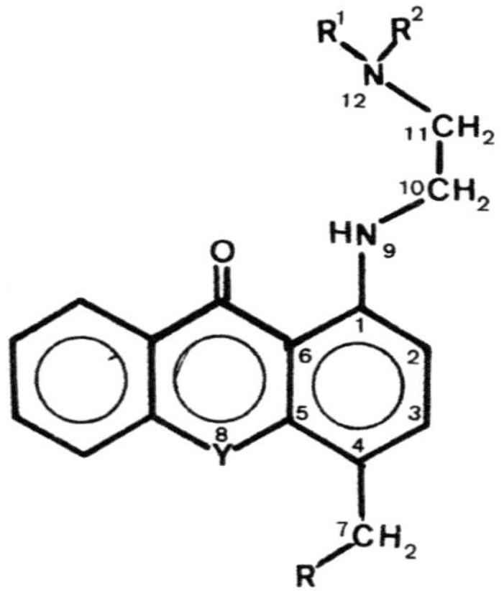

(I) $R^1, R^2 = C_2H_5$, Y = S, R = H
**Lucanthone**
(V) $R^1, R^2 = C_2H_5$, Y = S, R = OH
**Hycanthone**

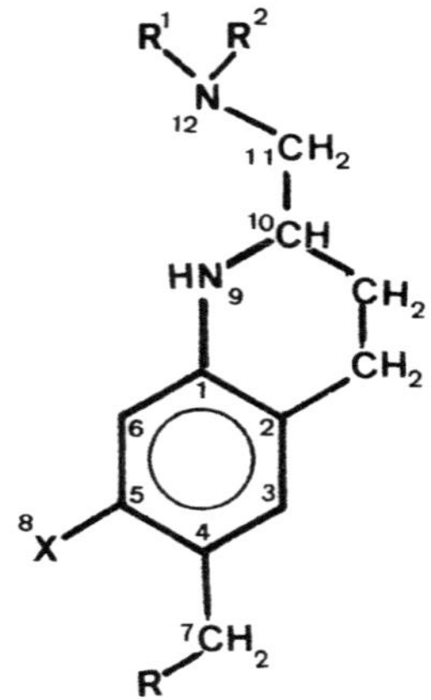

(II) R = H
(VI) R = OH

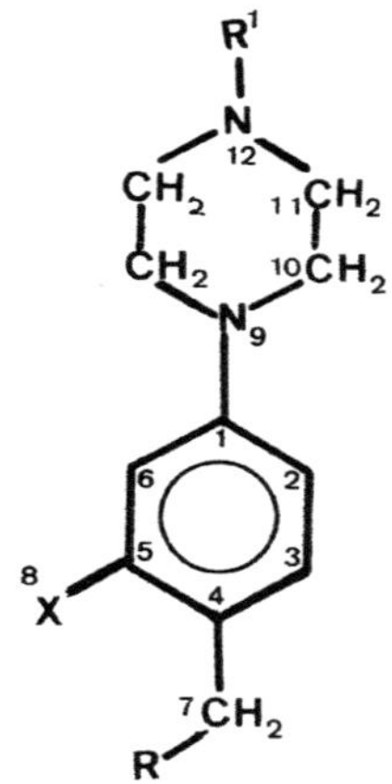

(III) R = H
(VII) R = OH

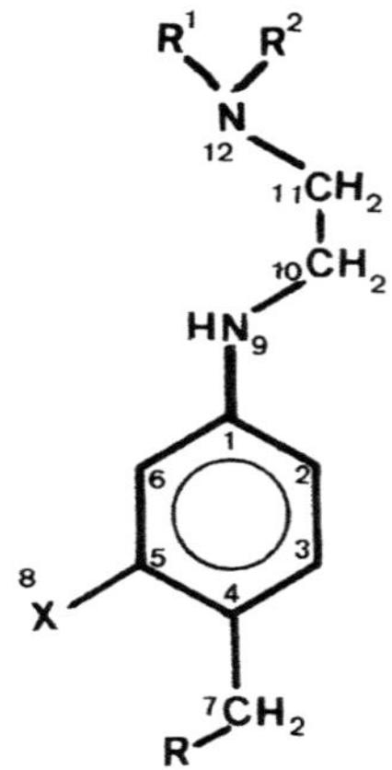

(IV) R = H
(VIII) = OH

Fig. 1. The molecular nucleus of some analogs and derivatives of Lucanthone ; R = hydrogen, hydroxyl ; $R^1$ = hydrogen, alkyl ; $R^2$ = alkyl ; Y = oxygen, sulfur ; X = halogen, cyano, nitro.

In order to establish such a fact unequivocally, more information is required than is currently available. In the present report we give the first results of a study undertaken in order to provide further testing of this theory, resting on evidence enterely different from the summarized experimental data.

On the hypothesis of a common mode of action of the compounds under consideration, the biological effect should be dependent upon the possibility that, on account of unfavorable bonded or nonbonded interactions within the molecules in the series, appropriate functional groups may assume the same unique biologically-significant stereochemical and electronic structure constituting the specific pharmocophore responsible for binding the schistosomal receptor site. To the extent that this holds true, the similarities in energetically preferential conformations throughout the molecular series as determined from quantum mechanical studies can provide relevant information for more detailed considerations of the structural basis of the schistosomicidal activity of these drugs. This paper describes a comparative study of the conformational possibilities and electronic structure of the representative nucleus of phenylpiperazine, aminomethyltetrahydroquinoline, aminoethylaminobenzene and aminoethylaminothioxanthenone series.

## II. Results

The comparative study of the conformational possibilities for the compounds under consideration concerns the spatial orientation with respect to the aromatic system of both the ethylenediamine chain and the hydroxymethyl para-substituent. The CNDO/2 semiempirical molecular orbital technique was used for these purposes. With a clear understanding of the well documented[20] capabilities and limitations of the method (in particular underestimation of internal barriers to rotation), it should be possible in principle to compute potential surfaces reasonably well accommodated to a comparative analysis of the conformational possibilities of molecules considered herein. The molecules were considered both isolated and in aqueous solution, the solvent effect upon the electronic molecular wave function was represented by the virtual charge (solvaton) model as recently revised and reformulated to include this effect within the CNDO/2 framework scheme(21). The molecular geometries used as imput data for the computations are taken from crystal structure data (22-25), and standard bond lengths and bond angles where no experimental data are available. In a first approximation, the conformational possibilities of the hydroxymethyl para-substituent could be considered a priori the same for all series of compounds in consideration, independently of the ethylenediamine chain orientation. These conformational possibilities were indeed determined from the 1.amino-4.hydroxymethyl-5 chlorobenzene derivative (with respect to the numbering system given in Figure 1).

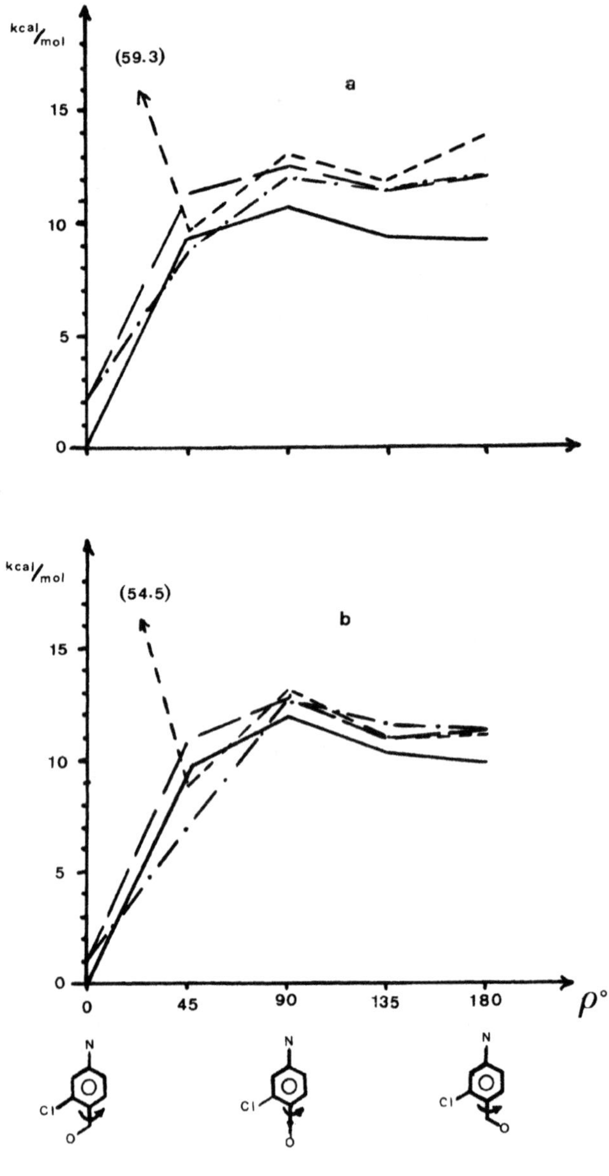


Fig. 2. CNDO/2 claculated energy *vs.* ρ($C_5$-$C_4$-$C_7$-O) for 1.amino-4. hydroxymethyl-5.chlorobenzene. (a) isolated ; (b) in aqueous solution. Energy plots : (———) ρ($C_4$-$C_7$-O-H)=180°, (— · —), ρ($C_4$-$C_7$-O-H)=60°, (—— ——) ρ($C_4$-$C_7$-O-H)=-60°, (- - - -) ρ($C_4$-$C_7$-O-H)=0°.

The 4-hydroxymethyl group orientation was treated as a two-rotomer problem, with the dihedral angle[+] $\rho(C_5-C_4-C_7-O)$ defining the deviation of the C-O bond from the plane of the aromatic ring and the dihedral angle $\rho(C_4-C_7-O-H)$ defining the orientation of the O-H bond about the C-O axis. Figure 2 illustrates the potential energy curves as a function of the dihedral angle $\rho(C_5-C_4-C_7-O)$, at dihedral $\rho(C_4-C_7-O-H)$ values 0°, 60°, -60° and 180°. It can be seen that the presence of the electronegative substituent ortho to the hydroxymetyl group confers a great conformational stability to the synplanar configuration $\rho(C_5-C_4-C_7-O)=0°$, but for $\rho(C_4-C_7-O-H)=0°$ with nonbonded atoms Cl and H within van der Waals contact distance. The mutual energy differences between the three staggered O-H configurations $\rho(C_4-C_7-O-H)=60°$, -60°, 180° are within 1 or 2 Kcal/mol in solution, the antiplanar $\rho(C_4-C_7-O-H)=180°$ form being in general the most stable. The only appreciable solvation effect seems to be the attenuation of nonbonded atoms interactions. In the subsequent analysis of the ethylenediamine chain configuration the rotomers $\rho(C_5-C_4-C_7-O)$ and $\rho(C_4-C_7-O-H)$ have been held fixed at $\rho(C_5-C_4-C_7-O)=0°$, $\rho(C_4-C_7-O-H)=180°$.

Neglecting the orientation of the $N_{12}$-alkylsubstituents, which have a large degree of conformational lability even in the solid state as instanced by their large thermal motion/disorder[22], the spatial orientation of the ethylenediamine chain in these compounds is essentially a three parameter problem. The parameters are the torsion angles $\rho(C_2-C_1-N_9-C_{10})$, $\rho(C_1-N_9-C_{10}-C_{11})$ and $\rho(N_9-C_{10}-C_{11}-\overset{+}{N}_{12})$. Nevertheless, the particular type of conformational restrictions, compatible with the schistosomicidal activity, imposed by the nucleus structure of N-phenylpiperazine and tetrahydroquinoline series greatly simplifies the search for a possible common pharmacophoric molecular conformation for these substances.

N-Phenylpiperazine nucleus. The ethylenediamine chain in this series is a part of the piperazine cycle. The conformational study reduces here to a one-parameter problem with constrained $\rho(C_1-N_9-C_{10}-C_{11})$ and $\rho(N_9-C_{10}-C_{11}-\overset{+}{N}_{12})$ coordinates. The nitrogen atoms are necessarily synclinal to each other, with a torsion angle $\rho(N_9-C_{10}-C_{11}-\overset{+}{N}_{12})$ of 53.7° due to the chair conformation of the piperazinium ring, and the torsion angle $\rho(C_1-N_9-C_{10}-C_{11})$ is antiplanar with a value of 180° due to the equatorial position of the benzene ring and the tetrahedral nature of the atom $N_7$. The structure of this series nucleus is similar to the structure of 1,1-dimethyl-4-phenylpiperazinium[25] (X), a potent

(X) 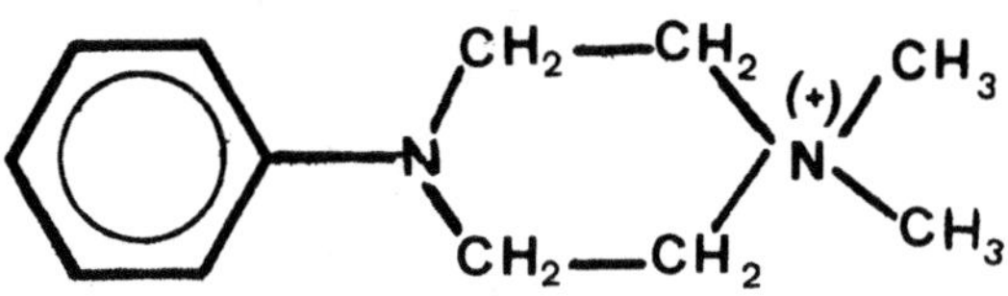


+ The dihedral angles are defined for the sequence A-B-C-D for a clockwise rotation of C-D relative to A-B when viewed along the bond B-C, with 0° as the synplanar conformation.

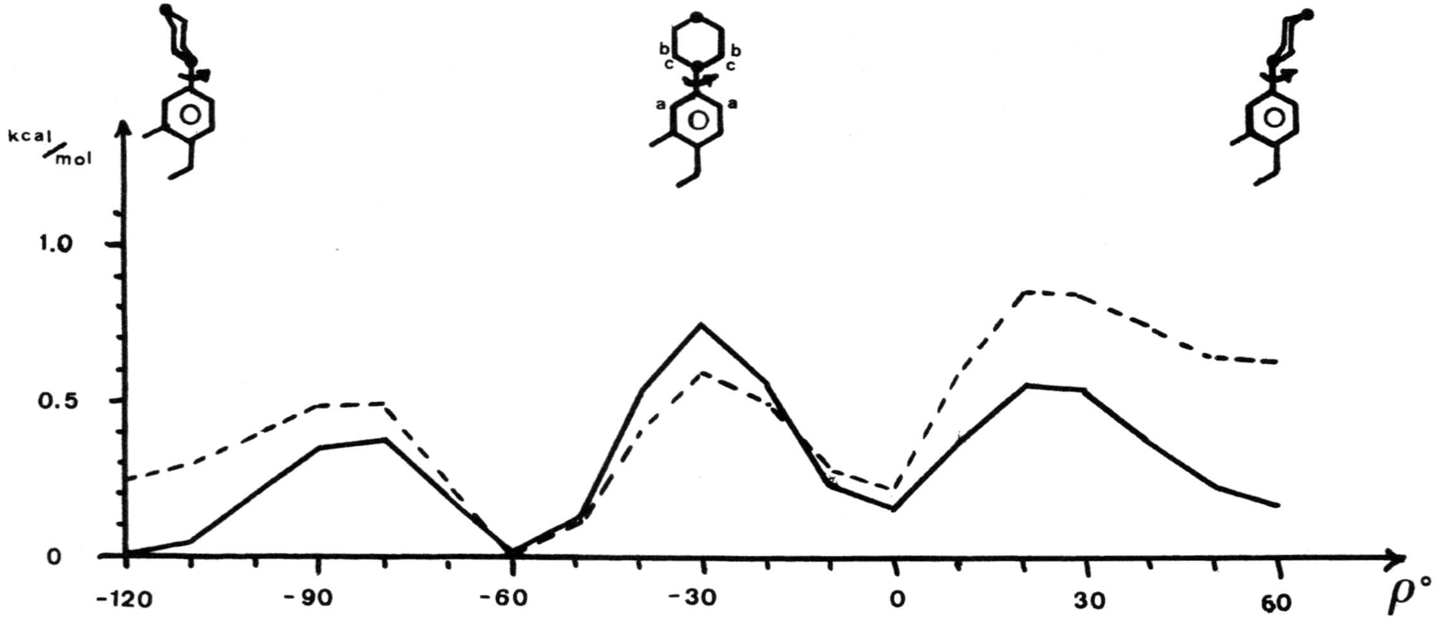


Fig. 3. CNDO/2 calculated energy *vs.* $\rho(C_2\text{-}C_1\text{-}N_9\text{-}C_{10})$ for N-(4.hydroxymethylphenyl-5.chloro)-piperazinium. Energy plots : (——) isolated, (- - -) in aqueous solution.

cholinergic stimulant in mammalian organisms.

The CNDO/2 conformational energies for the isolated and soluted molecule of N-(5.chloro-4.hydroxymethylphenyl)-piperazinium (VII) are plotted as a function of $\rho(C_2-C_1-N_9-C_{10})$ in Figure 3. The low computed barriers at $\rho(C_2-C_1-N_9-C_{10})=25°$, $-30°$, $-85°$ are due to conformations sterically restricted, with hydrogen atoms $H_a$ and $H_{b/c}$ within van der Waals contact distances. The shape of the curves indicates considerable lability in $\rho(C_2-C_1-N_9-C_{10})$. None of the conformations on the $\rho(C_2-C_1-N_9-C_{10})$ potential surface are energetically preclude as it may be confirmed by the crystal geometry $\rho(C_2-C_1-N_9-C_{10})=-30°$ observed for the compound (X)[25]. Nevertheless, it is useful to remark that the local minimum value $\rho(C_2-C_1-N_9-C_{10})=0°$, together with rotation angles $\rho(C_1-N_9-C_{10}-C_{11})=180°$ and $\rho(N_9-C_{10}-C_{11}-\overset{+}{N}_{12})=53.7°$ forced by the piperazinium ring, pertain to a spatial orientation of the $-N-C-C-\overset{+}{N}-$ chain approaching the $\{0°,\sim180°,\sim85°\}$ and $\{0°,\sim180°,\sim75°\}$ geometries implicated respectively in ACh muscarinic and nicotinic actions in mammals[(26-28)].

Aminomethyltetrahydroquinoline series nucleus.- In this series of compounds the conformation of the saturated heterocycle is that of a distorted semi-chair, practically a bent-chair[(24)], with $C_{14}$ located a distance of 0.772 Å above the aromatic plane, an equatorial aminomethyl-substituent in the 10-position, and constrained $\rho(C_2-C_1-N_9-C_{10})=0°$ and $\rho(C_1-N_9-C_{10}-C_{11})=208°$ coordinates. Assuming a staggered configuration for the terminal nitrogen group, the calculated CNDO/2 conformational energies for the isolated and soluted molecule are plotted as a function of $\rho(N_9-C_{10}-C_{11}-\overset{+}{N}_{12})$ in Figure 4. The calculation show three energy minima at the staggered configurations $\rho(N_9-C_{10}-C_{11}-\overset{+}{N}_{12})=60°$, $-60°$ and $180°$ which are separated by computed substantial barriers of some 4.5 Kcal/mol. The antiplanar conformation is computed to be about 1 Kcal/mol higher than the two synclinal forms. The only noticeable computed effect of solvation seems to be the attenuation of $N_9---\overset{+}{N}_{12}$ electrostatic interaction as substantiated by a lowering of about 1 Kcal/mol of the barrier located at $\rho(N_9-C_{10}-C_{11}-\overset{+}{N}_{12})=0°$. Together with the coordinate values $\rho(C_2-C_1-N_9-C_{10})=0°$ and $\rho(C_1-N_9-C_{10}-C_{11})=208°$ constrained by the saturated heterocycle, the $\rho(N_9-C_{10}-C_{11}-\overset{+}{N}_{12})=60°$ energy minimum pertains to the $-N-C-C-\overset{+}{N}-$ geometry represented as $\{0°,208°,60°\}$ near the muscarinic and nicotinic active geometries of ACh.

Aminoethylaminobenzene series nucleus.- As it is the case for xanthenone and heteronaphthalene series, the configuration of the ethylenediamine chain in this series is not constrained by intramolecular ring structure. The insubstituted proximal nitrogen group should be planar in principle : the resonance energy and the interaction between benzene cycle H-2 and hydrogens atoms attached to $C_{10}$ tend to stabilize a synplanar configuration of the rotomer $\rho(C_2-C_1-N_9-C_{10})$. The calculated energies vs. $\rho(N_9-C_{10}-C_{11}-\overset{+}{N}_{12})$ with $\rho(C_1-N_9-C_{10}-C_{11})=180°$ and $208°$, assuming a staggered configuration for the terminal amonium group, is shown in Figure 5. On the top of the Figure, the calculated energy vs. $\rho(N_9-C_{10}-C_{11}-\overset{+}{N}_{12})$ with $\rho(C_1-C_9-C_{10}-C_{11})=180°$, assuming an eclipsed

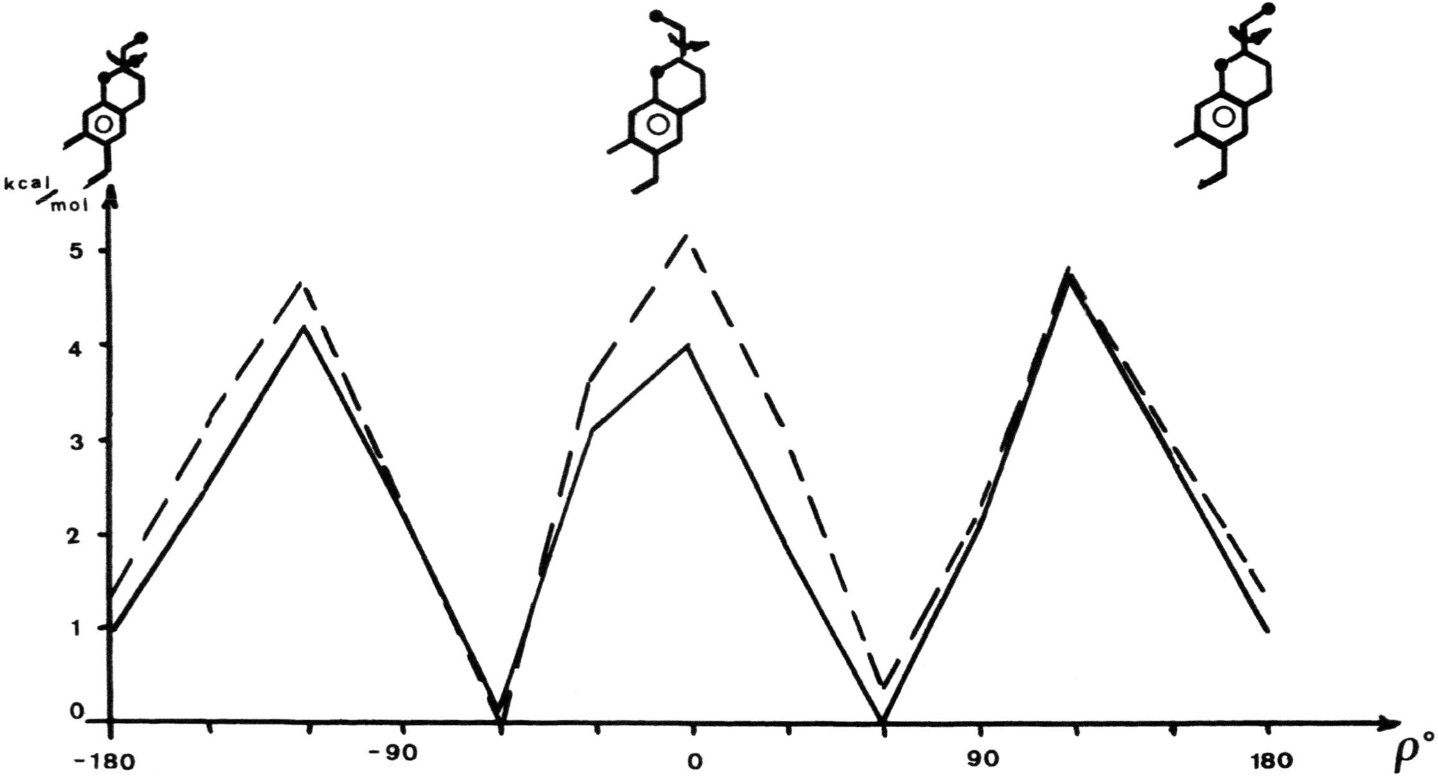


Fig. 4. CNDO/2 calculated energy *vs*. $\rho(N_9-C_{10}-C_{11}-\overset{+}{N}_{12})$ for 4.hydroxymethyl-5.chloro-10. aminomethyltetrahydroquinoline. Energy plots : (——) isolated, ( - - -) in aqueous solution.

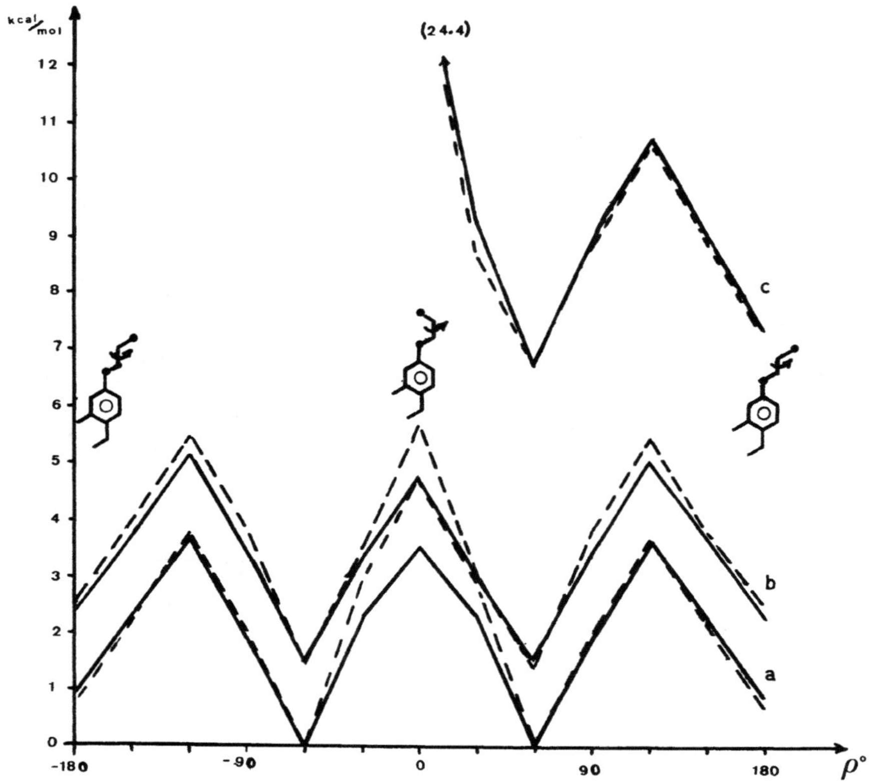


Fig. 5. CNDO/2 calculated energy *vs*. $\rho(N_9-C_{10}-C_{11}-\overset{+}{N}_{12})$ for 1.aminoethylamino-4.hydroximethyl-5.chlorobenzene. Staggered ammonium : (a) $\rho(C_1-N_9-C_{10}-C_{11})=180°$, (b) $\rho(C_1-N_9-C_{10}-C_{11})=208°$; eclipsed ammonium : (c) $\rho(C_1-N_9-C_{10}-C_{11})=180°$. Energy plots : (——) isolated, (- - -)in aqueous solution.

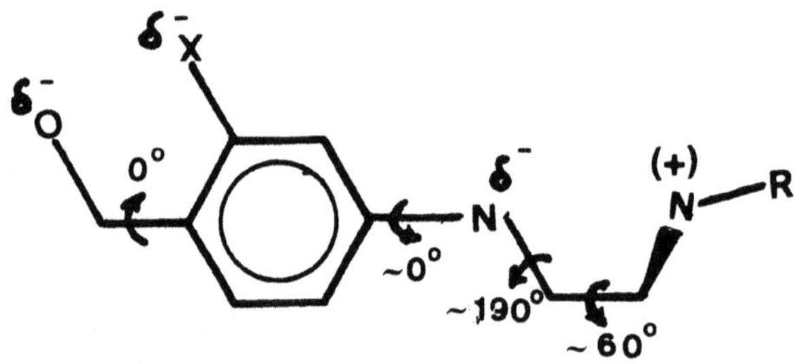


Fig. 6. Antischistosomal pharmacphoric pattern.

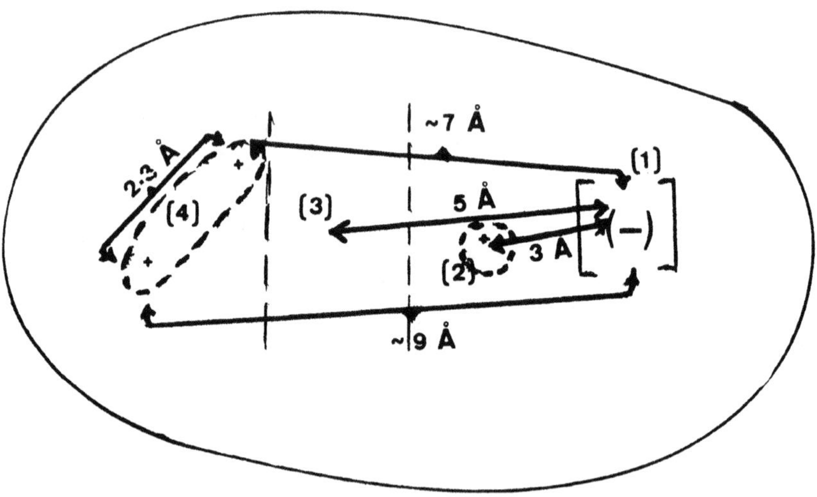


Fig. 7. Diagramm summarizing the essential features of the antichistosomal receptor site.

configuration for the terminal group as may be forced by the presence of two bulky alkyl-substituents at the nitrogen, is also represented. The calculated energy minima correspond to $\rho(N_9-C_{10}-C_{11}-\overset{+}{N}_{12})$=60°, - 60° and 180°, with the antiplanar form calculated to be 1 Kcal/mol higher than the synclinal forms. The eclipsed configuration at the terminal group leads to a substantial raise of the computed energies, but the position of the minima are essentially unchanged. As for the other compounds series, the only computed effect of solvation seems to be the attenuation of $N_9$---$\overset{+}{N}_{12}$ electrostatic interaction.

Xanthenone series nucleus.- The calculation on Lucanthone reported by other investigators(14), with the MINDO/3 approximation to molecular orbital theory, gives for the free molecule a global minimum whith the geometry {0°, 197.8°,76.5°} for the $-NH-CH_2-CH_2-\overset{+}{N}H-(CH_2-CH_3)_2$ side-chain.

## III. Discussion and conclusions.

To the extent specific pharmacophore is a dominant factor, compounds in these series with an assumed common mode of action by a blockade of schistosome ACh binding sites should have the appropriate functional groups disposed in three dimensional space in a common pattern. In terms of structural parameters, this disposition is specified by sterically permited or energetically preferred values of $\rho(C_2-C_1-N_9-C_{10})$, $\rho(C_1-N_9-C_{10}-C_{11})$, $\rho(N_9-C_{10}-C_{11}-\overset{+}{N}_{12})$ and $\rho(C_3-C_4-C_7-O)$, and a common pattern should be revealed by a comparison of allowed regions or energy minima in these parameters conformational maps. The conformational data presented in the preceding paragraphs have been reduced to tabular form and collected with the ACh cholinergic activity data in Table I. Each series nucleus is presented by a row of the Table, containing from left to right common-pattern preferred conformations of the ethylenediamine chain, corresponding intermolecular atomic distances and calculated distribution of electronic charges. In examination of the data it appears that in fact there is a $-N-C-C-\overset{+}{N}-$ {synplanar,antiplanar, synclinal} preferred geometry which is common to flexible and rigidified hycanthone analogs and derivatives. This preferred geometry is closely related to ACh proposed biologically active geometry of the choline chain $-O-C-C-\overset{+}{N}-$.

These results support the hypothesis that the ACh binding sites blockade is the common mechanism of action responsible for the schistosomicidal efficacy of these groups of compounds. Together with the experimental observations discussed above, they suggest that the activity of these drugs may be explained by the common pharmacaphoric pattern in Figure 6 wich embodies the specific structural and conformational characteristics required for full antischistosomal activity. The differences in potency and behaviour of these drugs should depend on the nature of the $\overset{+}{N}$-alkyl and 5-electronegative substituent and on several other particular molecular features such as the nature of the aromatic system, substitutions at the aromatic ring and at the proximal nitrogen atom, modifications of the ethylene chain fragment and so on. The schistosomicidal effects stems from the pharmacophoric pattern and charge distribution complementing those of the neighboring atoms in the receptor site.

The results may be resummed in the form of a diagram of the receptor site (Figure 7) which attempts to summarize the essential features probably responsible for binding these type of schistosomicidal drugs. In Figure 7 the distance between the binding points correspond approximately to the distances between the functional groups in the pharmacophoric pattern. This picture is more a summary of data than a map of the receptor site. The important points of receptor attachement are likely to be : (1) an anionic cavity negatively charged to accommodate the terminal nitrogen group of the aliphatic chain ; (2) a positively charged point accommodating the proximal nitrogen atom ; (3) a planar area for hydrophobic or van der Waal's binding of the aromatic system ; (4) a positively charged area to accommodate the hydroxylic and electronegative X-substituents. (1) and (2) indicated attachement points are common with generally acepted binding points in mammalian ACh muscarinic receptors[28], whereas (3) and (4) indicated attachement points may be considered as being probably responsible for the specific difference between schistosome and mammalian cholineceptive receptors suggested from experimental results[15-17].

On the whole, these observations point to a basis for speculation on the design of selective antischistosomal agents, selective for the worms with reduced or eliminated potential toxic host effects, while maintaining or increasing binding energy in anticipation of the synthesis of active compounds. Still refinement of both the experimental data and the theoretical calculations is required. These results must be considered in the context of data on a larger set of representative compounds, including refined conformational analysis. The charge distribution information should be produced by direct integration of electronic distribution over defined volumes or from electrostatic molecular potential fields, providing details of electronic topography rather than localized charge patterns. Work is currently in progress in our group on each of these points.

References.

1. S. Archer and A. Yarinsky, Progr. Drug. Res., 16, 11 (1972) and references therein.
2. P.B. Hulbert, E. Bueding and P.E. Hartman, Science 186, 647 (1974).
3. E. Bueding, J. Pharm. Pharmacol., 11, 385 (1959) and references therein.
4. W. Kikuth and R. Gönnert, Ann. Trop. Med. Parasitol., 42, 256 (1948).
5. R. Gönnert, Bull. World Health Org., 25, 702 (1961).
6. D. Rosi, T.R. Lewis, R. Lorenz, H. Freele, D.A. Berberian and S. Archer, J. Med. Chem., 10, 877 (1967).
7. E. Hirschberg (1974), in Antibiotics, Vol. 3, "Antimicrobial and Antitumor Agents" p. 274 (J.W. Corcoran and F.E. Hahn eds., Springer-Verlag, Berlin) and references therein.
8. H. Mauss. H. Kölling und R. Gönnert, Med. U. Chem., 5, 185 (1956) ; D. Rosi, G. Peruzzotti, E.W. Dennis, D.A. Berberian, H. Breele, B.F. Tullar and S. Archer, J. Med. Chem., 10, 867 (1967) ; D.A. Berberian, E.W. Dennis, H. Freele, D. Rosi, T.R. Lewis and S. Archer,

J. Med. Chem. 12, 607 (1969) ; C.A.R. Baxter and H. C. Richards, J. Med. Chem. 14, 1033 (1971).
9. G.R. Hillman, A.W. Senft and W.B.G. Gibler, J. Parasitol. 64, 754 (1978).
10. D. Cioli and P.M. Knopf, Am. J. Trop. Med. Hyg., 29, 220 (1980).
11. B. Weinstein and E. Hirschberg, Prog. Mol Subcell. Biol., 2; 232 (1971).
12. E. Bueding, J. Fisher and J. Bruce, J. Pharmacol. Exp. Therap., 186, 402 (1973).
13. M.H. Sarma, C.K. Mitra, R.H. Sarma and S. Archer, Int. J. Quant. Chem. : Quant. Biol. Symp., 7, 1 (1980).
14. K.J. Miller, M. Lauer and S. Archer, Int. J. Quant. Chem. : Quant. Biol. Symp., 7, 11 (1980).
15. L.R. Barker, E. Bueding and A.R. Timms, Brit. J. Pharmacol., 26, 656 (1966).
16. G.R. Hillman, W.B. Gibler and S.H. Chu, Biochem. Pharmacol., 25, 2529 (1976).
17. G.R. Hillman and A.W. Senft, Am. J. Trop. Med. Hyg., 24, 827 (1975).
18. B.J. Bogitsh, J. Parasitol., 61, 237 (1975).
19. A.W. Senft. D.G. Senft, G.R. Hillman, D. Polk and S. Kryger, Am. J. Trop. Med. Hyg., 25, 832 (1976).
20. J.A. Pople and D.L. Beveridge, "Approximate Molecular Orbital Theory", Academic, New York, (1972).
21. R. Constanciel and O. Tapia, Theoret. Chim. Acta (Berl.) 48, 75 (1978).
22. S. Neidle, Biochim. Biophys. Acta, 454, 207 (1976).
23. C.H. Wei and J.R. Einstein, Acta Cryst., B34, 205 (1978).
24. L.P. Zalukaev, N.A. Ignat'ev and E.I. Zavalishin, Zhumal. Struct. Khim., 16, 237 (1975).
25. C. Chothia and P. Pauling, Acta Cryst. B34, 2986 (1978).
26. C. Chothia and P. Pauling, Proc. Nat. Acad. Sci., 65, 477 (1970).
27. C. Chothia, Nature 225, 36 (1970).
28. C. Chothia, R.W. Baker and P. Pauling, J. Mol. Biol. 105, 517 (1976).

# INTERMOLECULAR INTERACTIONS AND SOLVENT EFFECTS : SIMPLIFIED THEORETICAL METHODS

Pierre CLAVERIE
Institut de Biologie Physico-Chimique,
Laboratoire de Biochimie Théorique
13, rue Pierre et Marie Curie - 75005 Paris.

ABSTRACT.

Two main topics are considered in this paper : the evaluation of the interaction energy between two molecules (through simplified analytical formulae) and the theoretical study of solvent effects through approximate methods (which make use of the previously described simplified interaction formulae). In section A, we briefly review the present state of the "exchange" or "symmetrized" perturbation theories of intermolecular interactions, which provide a convenient background for the derivation of simplified formulae. In section B, we recall the main features of the derivation of such formulae for the following interaction terms : electrostatic, short-range repulsion, polarization, dispersion, and we summarize some recent improvements. In section C, we first describe a theoretical framework for the treatment of solvent effects (definition of a cavitation term and a solute-solvent interaction term); then we recall the essential features of three models : discrete, continuum and mixed (discrete-continuum); we describe the statistical-mechanical foundations of the continuum model and the procedures for evaluating the corresponding interaction terms.

## A. THEORETICAL BACKBROUND.

Despite the advances in the field of ab initio (variational or perturbational) computations of molecular complexes, there is an obvious need for simplified formulae allowing the evaluation of intermolecular interaction energies with a computation time as short as possible : the use of such simplified procedures is necessary for making possible the treatment of systems involving either large molecules, or a large number of molecules (or both features simultaneously!).

In order to get such formulae we shall use here the perturbation theory of intermolecular interactions, because it allows us, in principle, to split the interaction energy into several terms, each



*R. Daudel, A. Pullman, L. Salem, and A. Viellard (eds.), Quantum Theory of Chemical Reactions, Volume III, 151–175.*

one of them having some rather simple variation law as a function of the intermolecular distance and the relative orientation of the molecules (more precisely, the various interatomic distances will be involved). It will then become easier to find separately for each term an appropriate analytic (approximate!) formula, and to get thus a simplified evaluation of the total interaction energy.

The total Hamiltonian H of the complex may be written as

$$H = H_0 + V \tag{A.1}$$

where $$H_0 = H^{(1)} + H^{(2)} \tag{A.2}$$

and $H^{(i)}$ denotes the Hamiltonian of the isolated molecule (i) (see e.g. [1], section II.A). The standard use of Rayleigh-Schrödinger (RS) perturbation theory correctly gives the asymptotic behaviour of the interaction energy at large distances (see e.g. [2]), but it fails to represent correctly this interaction at short distances (roughly speaking, the short-range repulsion is missing). A genuine understanding of this behaviour involves rather subtle topics, related with the increase of permutational symmetry (with respect to electron exchange) when going from $H_0$ to H, and significant progress concerning this subject occured essentially during the last decade [3-12 and references to previous works therein]. We shall not dwell here into a detailed theoretical study. We just recall that the usual RS treatment applied to $H_0$ as unperturbed Hamiltonian (with V as the perturbation operator) does not give an appropriate connection between the eigenstates of $H_0$ and H [3]. In order to overcome this defect, various so-called "exchange" or "symmetrized" perturbation treatments have been proposed [4-11], [12, section II-B and references therein]. The fact that the procedure is not unique may be easily understood as follows : we want to evaluate the energy difference $\Delta E = E - E_0$ between the (physical) ground state eigenvalue E of the hamiltonian H, defined by

$$(H-E)\ \Psi = 0 \tag{A.3}$$

and the corresponding eigenvalue $E_0$ of the hamiltonian $H_0$ corresponding to the non-interacting sub-systems) :

$$(H_0 - E_0)\ \Phi_0 = 0 \tag{A.4}$$

This problem may be treated by introducing an "interpolation" eigenvalue equation [13] :

$$F\left[\lambda\ ,E(\lambda)\right]\ \ \Phi\ (\lambda) = 0 \tag{A.5}$$

where $F(\lambda, E)$ is an "interpolation operator" connecting the two eigenvalue equations (A.3) and (A.4), i.e. an operator such that, for $\lambda = 0$, we have $E(0) = E_0$ (the eigenvalue defined by (A.4)), and for $\lambda = 1$ we have $E(1) = E$ (the eigenvalue defined by (A.3)). The requirements for $\Phi(\lambda)$ are not unique : usually, it will be required that $\Phi(0) =$

$\varphi_o = W_o^{(1)} W_o^{(2)}$ (the eigenfunction of $H_o$, product of the eigenfunctions $W_o^{(i)}$ of $H^{(i)}$), but as concerns $\Phi(1)$, at least two important possibilities exist : either to impose $\Phi(1) = \Psi$(the fully antisymmetrized eigenfunction of H in (A-3)), or to require only that $\Phi(1) = \varphi$ with $\mathcal{A}\varphi = \Psi$ , where $\mathcal{A}$ denotes some antisymmetrization operator (*) ; $\varphi$ is called a "primitive function" [ 9-11,14]. Then, a perturbation treatment may be devised by expressing the difference E(1)-E(0) through the Taylor expansion of $E(\lambda)$ - E(0), which provides a power series in terms of $\lambda$(hence the desired perturbation expansion by putting $\lambda = 1$). Obviously, we have an infinity of such perturbation expansions, corresponding to the infinity of analytic curves $E(\lambda)$ which connect the two prescribed values $E_o$ and E. At first sight, such a lack of uniqueness could look discouraging, but it must be remembered that we have one more requirement of a very practical nature, namely that the perturbation expansion should converge as rapidly as possible (especially at long- and medium-range for the kind of application we are interested in). Now, Jeziorski and Kolos [ 7], Jeziorski et al. [8] have rederived several "symmetrized perturbation treatments" (EL-HAV : Eisenschitz-London, Hirschfelder-Van der Avoird; MS-MA : Murrell-Shaw, Musher-Amos) and proposed new ones (JK : Jeziorski-Kolos, SRS : Symmetrized Rayleigh-Schrödinger) in a common framework (Bloch form of the Schrödinger equation, iterative process with various possibilities of inserting antisymmetrization operators, hence generating various treatments). Then the following conclusions may be drawn :

(1) The EL-HAV treatment, which involves the strongest "symmetry forcing" [7,8], is not convenient for long- and medium range interaction because, already at 2nd-order, it does not exhibit the correct asymptotic behaviour : the coefficient of the $1/R^6$ term is strongly underestimated [ 14,15][16, Appendix], [ 17, p. 126 and Appendix D], [ 18, Chap. II, section 3]. Kutzelnigg [ 9-11 ] accounted for this feature by showing that, when expressed as a "primitive function" treatment, the EL-HAV method corresponds to a primitive function which is <u>not genuinely primitive</u>.

(2) The other three treatments (JK, MS-MA, SRS) not only exhibit the correct asymptotic behaviour up to the 2nd-order perturbation term, but actually give <u>identical</u> 1st- and 2nd-order terms. This identity of course breaks down for the higher-order terms, but MS-MA and SRS still exhibit the same correct asymptotic long-range behaviour (of the form $C_n/R^n$) which is given by the usual (non symmetrized) Rayleigh-Schrödinger (RS) treatment [ 2 ]. A thorough study of the convergence properties of MS-MA and SRS expansion reveals non-trivial and very interesting properties [ 5,6], but, from the practical point of view, an essential conclusion is that the 1st- and 2nd-order perturbation terms common to the JK, MS-MA and SRS treatments provide a conve-

(*) Several operators, differing by a "normalization constant" may be used : see e.g. ref. [ 1 ] , appendix A.

nient starting point for a reasonable approximate evaluation of the interaction energy at long- and medium-range.

In actual fact, it is possible to go still further : other "symmetrized" or "exchange" treatments have been proposed (see e.g. [12, section II-B] or [18, ch. II, section 4], and other references therein), and they have, as a general rule, the same 1st-order term and the same long-range part (namely $\varepsilon_2^{RS}$) in the 2nd-order term as the three treatments above : namely, differences occur only in the so-called "exchange" part of this 2nd-order term. Moreover, explicit evaluations of this 2nd-order exchange term suggest that it should be rather small for interactions between non polar systems (for $H...H^+$ [4], for H...H [14], for He...He [20] and the more elaborate treatments [21-23] ; see also [19] and the discussion in [1, section III.A.1.b]). For interactions between polar molecules, and especially in hydrogen bonded complexes, it could be feared that the 2nd-order exchange term plays a more important role, but the computation of the interaction for the water dimer by Jeziorski and Van Hemert [24] sugest that this is not the case : they indeed obtain quite satisfactory results by computing $\varepsilon_1 = \varepsilon_1^{RS} + \varepsilon_1^{exch}$ and $\varepsilon_2^{RS}$ only (RS : Rayleigh-Schrödinger; exch: exchange). What appears essential is to account properly for the short-range behaviour of $\varepsilon_2^{RS}$, and this is achieved in practice by including charge transfer states in the basis used for the calculation of the 1st-order perturbed wave-function (indeed, this compensates for the fact that the bases used for each subsystem are unavoidably finite, i.e. not complete, and, in particular, would give a rather poor representation of the continuum states of each subsystem); but performing complete antisymmetrization of all the excited states (whether of charge-transfer type or not) would be much less important.

Accordingly, we shall follow the proposal already made by several authors [25, 18, 1, 24], namely to evaluate the interaction energy as

$$\Delta E \simeq \varepsilon_1^{HL} + \varepsilon_2^{RS} \tag{A.6}$$

where

$$\varepsilon_1^{HL} = \varepsilon_1 = \varepsilon_1^{RS} + \varepsilon_1^{exch} \tag{A.7}$$

denotes the 1st-order term common to almost all exchange perturbation treatments proposed thus for (HL stands for Heitler-London : see explanation below). Whitton and Byers-Brown [26] consider $\varepsilon_1^{HL}$ as unsatisfactory because it exhibits an incorrect behaviour at very long distance (wrong sign for the H...H interaction); however, we are interested in the region of the Van der Waals minimum rather than in very large distances, and for our purpose the evaluation (A.6) seems quite sufficient. We shall now recall the expressions of the 1st- and 2nd-order perturbation terms (denoted $\varepsilon_1$ and $\varepsilon_2$ respectively), which are common to the three exchange perturbation treatments MS-MA [27,28], JK and SRS [7,8], as indicated above.

The eigenfunctions of $H^{(i)}$ (cf. eq. (A.2) will be denoted by :

$$W_0^{(i)},\ W_1^{(i)},\dots,W_a^{(i)},\dots,W_b^{(i)} \quad (A.8)$$

with the corresponding eigenvalues :

$$E_0^{(i)},\ E_1^{(i)},\dots,E_a^{(i)},\dots,E_b^{(i)},\dots \quad (A.9)$$

Then the eigenfunctions of $H_0 = H^{(1)} + H^{(2)}$ are simply the products :

$$W_0^{(1)} W_0^{(2)},\dots,W_0^{(1)} W_b^{(2)},\dots W_a^{(1)} W_0^{(2)},\dots,W_a^{(1)} W_b^{(2)},\dots \quad (A.10)$$

with the corresponding eigenvalues :

$$E_0^{(1)}+ E_0^{(2)},\dots,E_0^{(1)}+ E_b^{(2)},\dots,E_a^{(1)}+ E_0^{(2)},\dots,E_a^{(1)}+ E_b^{(2)},\dots \quad (A.11)$$

Besides the explicit notation (A.10), a more compact notation is also useful for the eigenfunctions and eigenvalues of $H_0$, namely :

$$\varphi_0\ (= W_0^{(1)} W_0^{(2)}),\ \varphi_1,\dots,\varphi_i,\dots \quad (A.12)$$

$$E_0^0\ (= E_0^{(1)}+ E_0^{(2)}),\ E_1^0,\dots,\ E_i^0,\dots \quad (A.13)$$

The expressions of $\varepsilon_1$ and $\varepsilon_2$ from ref. [ 7, 8, 27, 28] may then be written :

$$\varepsilon_1 = < \varphi_0 | V \mathcal{A} | \varphi_0 > / < \varphi_0 | \mathcal{A} | \varphi_0 > \quad (A.14)$$

$$\varepsilon_2 = - < \varphi_0 | V \mathcal{R}_0 \mathcal{A}(V-\varepsilon_1) | \varphi_0 > / < \varphi_0 | \mathcal{A} | \varphi_0 > \quad (A.15)$$

where $$\mathcal{R}_0 = \sum_k{}' \ | \varphi_k >< \varphi_k | / (E_k^0 - E_0^0) \quad (A.16)$$

is reduced resolvent of $H_0$ associated with the eigenfunction $\varphi_0$ (the prime in $\Sigma'$ means that $k = 0$ is excluded from the summation). It must be emphasized that $\varepsilon_1$ given by (A.14) is nothing but the "Heitler-London" approximation to the interaction energy (that we may denote $\varepsilon_1^{HL}$) namely the difference between the total Heitler-London energy $E^{HL} = < \Psi_0 | H | \Psi_0 > / < \Psi_0 | \Psi_0 >$ (where $\Psi_0 = \mathcal{A} \varphi_0$) and the unperturbed energy $E_0^0 =< \varphi_0 | H_0 | \varphi_0 >$ . Proof (see e.g. [1], section II.B.3.b, p.112) :

$$\frac{< \Psi_0 | H | \Psi_0 >}{< \Psi_0 | \Psi_0 >} = \frac{< \mathcal{A}\varphi_0 | H | \mathcal{A} \varphi_0 >}{< \mathcal{A}\varphi_0 | \mathcal{A}\varphi_0 >} + \frac{< \varphi_0 | H | \mathcal{A}^2 \varphi_0 >}{< \varphi_0 | \mathcal{A}^2 \varphi_0 >}$$

$$= \frac{< \varphi_0 | H_0 | \mathcal{A}^2 \varphi_0 >}{< \varphi_0 | \mathcal{A}^2 \varphi_0 >} + \frac{< \varphi_0 | V | \mathcal{A}^2 \varphi_0 >}{< \varphi_0 | \mathcal{A}^2 \varphi_0 >} \quad (A.17)$$

Since $\langle \varphi_0 | H_0 = E_0^0 \langle \varphi_0 |$, the first term reduces to $E_0^0$. In the second term, $\mathcal{A} = c\,A$ where A is the antisymmetrization projection operator (hence $A^2 = A$), therefore $\mathcal{A}^2 = c^2 A^2 = c^2 A = c\mathcal{A}$, and we may accordingly replace $\mathcal{A}^2$ by $\mathcal{A}$ in both the numerator and denominator, hence the desired result :

$$E^{HL} = E_0^0 + \langle \varphi_0 | V | \mathcal{A} \varphi_0 \rangle / \langle \varphi_0 | \mathcal{A} \varphi_0 \rangle = E_0^0 + \varepsilon_1 \tag{A.18}$$

Note that the peculiar value of the constant c appearing in the definition of $\mathcal{A}$ makes no difference as concerns $\varepsilon_1$ and $\varepsilon_2$, since $\mathcal{A}$ appears in both the numerator and the denominator. For the sake of definiteness, and in order to make easier the comparison with the usual RS treatment, we shall use (cf. [ 1], Appendix A) :

$$\mathcal{A} = 1 - \mathcal{A}' = 1 - P_{(1)} + P_{(2)} - \ldots + (-1)^{n_0} P_{(n_0)} \tag{A.19}$$

where $P_{(1)} = \sum_i^{(1)} \sum_j^{(2)} P_{ij}$ denotes the sum of all permutations which exchange 1 electron (i) of molecule (1) and 1 electron (j) of molecule (2), and similar definition for $P_{(2)}$, $P_{(3)} \ldots$ (of course, since this definition (A.19) involves only inter-system exchanges, the operator $\mathcal{A}$ is supposed to act over functions such as $W_a^{(1)} W_b^{(2)}$ which are already antisymmetric with respect to all intra-system exchanges).

Now, we shall explicitly decompose $\varepsilon_n$ into the usual RS perturbation term $\varepsilon_n^{RS}$, and a so-called exchange part $\varepsilon_n^{exch}$ :

$$\varepsilon_n = \varepsilon_n^{RS} + \varepsilon_n^{exch} \tag{A.20}$$

where the $\varepsilon_n^{RS}$ (n = 1,2) are given by the well-known formulae (see e.g. [29], vol. II, chap. XVI) :

$$\varepsilon_1^{RS} = \langle \varphi_0 | V | \varphi_0 \rangle \tag{A.21}$$

$$\varepsilon_2^{RS} = - \langle \varphi_0 | V \mathcal{R}_0 V | \varphi_0 \rangle = - \sum_k{}' \frac{|\langle \varphi_0 | V | \varphi_k \rangle|^2}{E_k^0 - E_0^0} \tag{A.22}$$

The decomposition (A.20) is easily performed by inserting the decomposition (A.19) of into (A.14) and (A.15) (see [1], section IV.A, p.138 eq. (2.158)) :

$$\varepsilon_1 = \frac{\langle \varphi_0 | V | (1-\mathcal{A}') \varphi_0 \rangle}{\langle \varphi_0 | \mathcal{A} \varphi_0 \rangle} = \frac{\langle \varphi_0 | V | \varphi_0 \rangle}{\langle \varphi_0 | \mathcal{A} | \varphi_0 \rangle} - \frac{\langle \varphi_0 | V \mathcal{A}' | \varphi_0 \rangle}{\langle \varphi_0 | \mathcal{A} | \varphi_0 \rangle}$$

and

$$\frac{1}{\langle \varphi_0 | \mathcal{A} | \varphi_0 \rangle} = 1 - \frac{\langle \varphi_0 | \mathcal{A} | \varphi_0 \rangle - 1}{\langle \varphi_0 | \mathcal{A} | \varphi_0 \rangle} = 1 + \frac{\langle \varphi_0 | \mathcal{A}' | \varphi_0 \rangle}{\langle \varphi_0 | \mathcal{A} | \varphi_0 \rangle}$$

since $< \varphi_0 |\mathcal{A}| \varphi_0 > = < \varphi_0| 1-\mathcal{A}'|\varphi_0 > = 1 - < \varphi_0| \mathcal{A}'| \varphi_0 >$ , hence

$$\varepsilon_1 = < \varphi_0| V| \varphi_0 > + \frac{<\varphi_0| V|\varphi_0 > < \varphi_0|\mathcal{A}'|\varphi_0 > - < \varphi_0| V\mathcal{A}'|\varphi_0 >}{< \varphi_0 |\mathcal{A}| \varphi_0 >} \tag{A.23}$$

where the first term is $\varepsilon_1^{RS}$, therefore the second is $\varepsilon_1^{exch}$. The numerator of $\varepsilon_1^{exch}$ actually involves inter-system electron exchanges, and therefore it decreases exponentially as a function of intermolecular distance. On the contrary, the denominator $<\varphi_0|\mathcal{A}| \varphi_0> = 1 - < \varphi_0 |\mathcal{A}'| \varphi_0 >$ could be replaced by 1 as a first approximation (retaining the main terms of order $S^2$ in $\varepsilon_1^{exch}$, where S denotes the order of magnitude of the intermolecular overlap integrals between molecular orbitals of the two subsystems).

Proceeding in the same way, we obtain for $\varepsilon_2$ :

$$\varepsilon_2 = - < \varphi_0| V \mathcal{R}_0 V |\varphi_0 > +$$

$$\Big[ < \varphi_0| V\mathcal{R}_0 \mathcal{A}'(V- \varepsilon_1)|\varphi_0 > - < \varphi_0| V\mathcal{R}_0 V|\varphi_0 > < \varphi_0|\mathcal{A}'| \varphi_0 >\Big]$$

$$/ < \varphi_0 |\mathcal{A}| \varphi_0 > \tag{A.24}$$

where we used the fact that $<\varphi_0| V\mathcal{R}_0 |(V-\varepsilon_1) \varphi_0 > = < \varphi_0| V\mathcal{R}_0 V|\varphi_0 > - \varepsilon_1 < \varphi_0| V\mathcal{R}_0|\varphi_0 > = < \varphi_0| V\mathcal{R}_0 V| \varphi_0 >$ since $\mathcal{R}_0 \varphi_0 = 0$. The first term of (A.24) is $\varepsilon_2^{RS}$, hence the second is $\varepsilon_2^{exch}$. Here too, we could, as a first approximation, replace $<\varphi_0|\mathcal{A}|\varphi_0 >$ by 1 in the denominator of $\varepsilon_2^{exch}$

The formulae (A.23,24) will be taken as the theoretical basis for deriving simplified approximate formulae.

## B. SIMPLIFIED FORMULAE FOR THE INTERACTION ENERGY

### 1. Some basic elements

As explained in the previous section, we shall consider the following approximation of the interaction energy :

$$\Delta E \simeq (\varepsilon_1^{RS} + \varepsilon_1^{exch}) + \varepsilon_2^{RS} \tag{B.1}$$

i.e. we discard $\varepsilon_2^{exch}$ and the perturbation terms beyond 2nd-order. By introducing the more explicit form (A.10,11) of the eigenfunctions and eigenvalues of $H_0$, the expressions (A.21,22) of $\varepsilon_1^{RS}$ and $\varepsilon_2^{RS}$ become (see e.g. [ 1] , section II.A.1, or [ 30], section 3) :

$$\varepsilon_1^{RS} = \left\langle W_0^{(1)} W_0^{(2)} | V | W_0^{(1)} W_0^{(2)} \right\rangle \tag{B.2}$$

$$\varepsilon_2^{RS} = - \sum_a' \frac{|< W_0^{(1)} W_0^{(2)} | V | W_a^{(1)} W_0^{(2)} >|^2}{E_b^{(1)} - E_0^{(1)}} - \sum_b' \frac{|< W_0^{(1)} W_0^{(2)} | V | W_0^{(1)} W_b^{(2)} >|^2}{E_b^{(2)} - E_0^{(2)}} - \sum_a' \sum_b' \frac{|< W_0^{(1)} W_0^{(2)} | V | W_a^{(1)} W_b^{(2)} >|^2}{(E_a^{(1)} - E_0^{(1)}) + (E_b^{(2)} - E_0^{(2)})} \tag{B.3}$$

As is well known, $\varepsilon_1^{RS}$ is the electrostatic interaction energy between the unperturbed charge distributions of the two molecules, the 1st term of (B.3) is the polarization (or induction) energy of molecule 1 by molecule 2 (this term indeed corresponds to the appearance of components $W_a^{(1)}$ in the perturbed function under the effect of the electric field of molecule 2), the 2nd term of (B.3) is the symmetric term (polarization of molecule 2 by molecule 2), and the 3rd term of (B.3) is the dispersion energy (see e.g. [1,30,31,33]). A basic tool for treating these terms (and, noticeably, obtaining the above-mentioned interpretations) is the expression of any matrix element of the interaction operator V as a classical electrostatic interaction energy between suitable charge distributions [31] :

$$< W^{(1)} W^{(2)} | V | W^{(1)} W^{(2)} > = \iint \frac{f_{ac}^{(1)}(\vec{r}_1)\, f_{bd}^{(2)}(\vec{r}_2)}{|\vec{r}_1 - \vec{r}_2|}\, d\vec{r}_1 d\vec{r}_2 \tag{B.4}$$

where

$$f_{ac}^{(1)}(\vec{r}) = < W_a^{(1)} | \rho^{(1)}(\vec{r}) | W_c^{(1)} > \tag{B.5a}$$

$$f_{bd}^{(2)}(\vec{r}) = < W^{(2)} | \rho^{(2)}(\vec{r}) | W_d^{(2)} > \tag{B.5b}$$

and $\rho$ $(\vec{r})$ is the "charge density operator" associated with molecule (m) :

$$\rho^{(m)}(\vec{r}) = \sum_\mu^{(m)} Z_\mu \delta\, (\vec{r} - \vec{r}_\mu) - \sum_i^{(m)} \delta(\vec{r} - \vec{r}_i) \tag{B.6}$$

refers to nuclei and i to the electrons ($Z_\mu$ denotes the charge of nucleus $\mu$). As a straightforward application, we see that $\varepsilon_1^{RS}$ , as given by (B.2), is actually equal to the electrostatic interaction between $f_{00}^{(1)}(\vec{r}_1)$ and $f_{00}^{(2)}(\vec{r}_2)$, which are the charge densities of molecule (1) and (2) in their respective ground states. The essential advantage of this representation (B.4) of the matrix elements is the following : they introduce quantities which depend only on each isolated molecule (namely the charge distributions $f_{ac}^{(1)}(r_1)$ and $f_{bd}^{(2)}(r_2)$), and the intermolecular interaction terms are then expressed as relatively simple functions (or functionals) of these purely molecular quantities which can, in principle, be tabulated once for all for every molecule, independently of its future partners. The interpretation of $\varepsilon_1^{RS}$ is typical in this respect, and further applications to the treatment of the 2nd-order terms (polarization and dispersion) have been developed in previous works [1, sections III to V], [30] : essentially, these terms are (approximately) decomposed into sums of bond polarization energies and sum of bond-bond dispersion energies, through the introduction of bond polarizabilities, both static and dynamic. The dispersion terms, in particular, are conveniently expressed as integrals over the dynamic (frequency-dependent) polarizabilities (taken for imaginary argument) see e.g. the review by Dalgarno and Davison [33] or [1, section V.E.3], and [32].

As concerns the short-range term $\varepsilon_1^{exch}$, it is also possible to develop similar treatments starting from the theoretical expression (essentially the numerator of the second term in (A.23)). For interaction between atoms (rare gases), such developments suggest the approximate expression [34], [1, section IV.C.2, eq. (2.195)]:

$$\varepsilon_1 \simeq \frac{C}{R} \sum_i^{(1)} \sum_j^{(2)} n_i^{occ}\, n_j^{occ}\, (S_{ij})^2 \qquad (B.7)$$

where R is the interatomic distance, i and j refer to the orbitals, $n_i^{occ}$, $n_j^{occ}$ denote their occupation numbers and $S_{ij}$ denotes the overlap integral. In (B.7) we consider $\varepsilon_1$ instead of $\varepsilon_1^{exch}$ because, for rare gases, the electrostatic energy $\varepsilon_1^{RS}$ reduces to the so-called "penetration" part (because all multipoles are exaclty zero), and this part behaves as an exponential function of R, as $\varepsilon_1^{exch}$ does, so that it is possible to represent their sum rather than $\varepsilon_1^{exch}$ only. For molecules, we may search for an approximate expression as a sum of local contributions [30], [1, section IV.D.E], e.g., as a generalization of (B.7), a sum of $(S_{ij})^2$ where i and j would be localized molecular orbitals. A practical application deriving from this proposal will be indicated below.

The accuracy of simplified formulae may be surprisingly good. We shall recall here two examples taken from [1].

(a) The 1st-order (Heitler-London) short-range interaction for two Helium atoms [1, section IV.B]. In such a simple case, it is possible to work out rather elaborate analytical formulae by using the explicit expressions of the integrals involved, and a good representation of $\varepsilon_1^{HL}$ may be obtained in that way. But a simple exponential fit may also be tried (by searching a linear representation of Log $\varepsilon_1$ through least squares fitting) :

$$\varepsilon_1 \simeq B \exp(-bR) \qquad (B.8)$$

and this representation appears quite satisfactory : two approximations of the Helium atomic orbital were considered, the first one made up from a single Slater 1s orbital ( $\zeta = 1.6875$), and the second one from four Slater orbitals (1s(1.4), 2s(1.4), 1s(3.0), 2s(3.0)). Then, according to tables 2.1 and 2.2 of ref. [ 1, section IV.B], the largest error (over the interval 4.5 u.a. < R < 7.0 u.a.) amounts to -5.61 % in the first case (1 component orbital), and to -2.65% in the second case (4 component orbital). It is quite remarkable that the smallest error is obtained in the second case, despite the fact that the exact analytical expression of $\varepsilon_1$ is much more cumbersome precisely in this case! For other rare gases, too (Neon, Argon), it appears possible to represent with sufficient accuracy the short-range part of the interaction energy through a simple exponential fit, at least in the region of the Van der Waals minimum [ 35, 36].

(b) if we consider the 1/R expansion of the dispersion energy between two (rare gas) atoms :

$$E_{disp} = - C_6/R^6 + C_8/R^8 + \dots \tag{B.9}$$

the $C_6$ coefficient may be expressed exactly as (see e.g. [ 33]) :

$$C_6 = \frac{3}{\pi} \int_{-\infty}^{\infty} \alpha^{(1)}(i\xi) \;\; \alpha^{(2)}(i\xi) \; d\xi \tag{B.10}$$

where $\alpha^{(m)}(i\xi)$ denotes the dynamic polarizability analytically continued along the imaginary axis, namely, for atoms [ 33, section II.C] :

$$\alpha(i\xi) = \frac{2}{3} \sum_I{}' \frac{\Delta E_I (\vec{\mu}_{0I})^2}{(\Delta E_I)^2 + \xi^2} \tag{B.11}$$

where 0 refers to the ground state, I to an excited state, $\Delta E_I$ is the transition energy, and $\vec{\mu}_{0I}$ the transition moment. Approximate formulae for $C_6$ may be derived by using for $\alpha(i\xi)$ some approximate expression instead of (B.11), e.g. a "one-term" formula :

$$\alpha(i\xi) \simeq \frac{\alpha}{1+(b\xi)^2} \tag{B.12}$$

where $\alpha = \alpha(0)$ is the static polarizability, and b is some suitable parameter (with dimension $(\text{Energy})^{-1}$). Upon inserting (B.12) into (B.10), we get :

$$C_6 \simeq \frac{3}{2} \frac{\alpha^{(1)} \alpha^{(2)}}{b^{(1)} + b^{(2)}} \tag{B.13}$$

which has the form of the usual London formula :

$C_6 = \frac{3}{2} \frac{U^{(1)} U^{(2)}}{U^{(1)}+U^{(2)}} \alpha^{(1)} \alpha^{(2)}$, provided we take $U^{(1)} = 1/b^{(1)}$ and $U^{(2)} = 1/b^{(2)}$. Now, in [1, section V.E.3], two choices for the parameter b are discussed : Mavroyannis-Stephen [ 37] (Ma-St), and Salem [ 38] -Tang-Karplus [ 39] (STK) :

$$b(\text{Ma-St}) = \sqrt{\alpha}/n \qquad (B.14)$$

$$b(\text{STK}) = \frac{3}{2} \alpha/(\Delta\vec{\mu})^2 \qquad (B.15)$$

In (B.14), n denotes the number of the electrons of the atom, and in (B.15), $(\Delta\vec{\mu})^2 = \langle 0|(\vec{\mu} - \langle\vec{\mu}\rangle)^2|0\rangle (= \langle 0|(\vec{\mu})^2|0\rangle$ since $\langle \vec{\mu} \rangle = 0$ for an atom, owing to spherical symmetry) denotes the dispersion of the dipole moment operator $\vec{\mu} = e \sum_i \vec{r}_i$). Now, upon comparison of these approximate $C_6$ with reasonable estimates of the exact ones (see e.g. [ 40 ], table V), the following conclusions appear :

(1) the Ma-St choice leads to overestimated values (see e.g. [ 41], table II), so that, in order to get reasonable values, it is necessary to introduce "effective electron numbers" smaller than the true ones [ 41].

(2) by contrast, the STK choice leads to quite accurate values of $C_6$ without any ad hoc modification of the quantities $\alpha$ and $(\Delta\vec{\mu})^2$ (see [ 1], section V.E.3.b, table 2.4, p. 203). It is quite remarkable that the values are uniformly good for all possible pairs of rare gases from Helium to Xenon (whether homoatomic or heteroatomic pairs).

Thus, in both cases (a) (short-range repulsion) and (b) ($C_6/R^6$ dispersion term), we have found it possible to get a sufficiently accurate representation through rather simple formulae. Of course, we were dealing with interactions between atoms only, but such results are nevertheless quite encouraging.

2. Simplified formulae for the interaction between molecules.

We have just described some examples of rather successful simplified representations of interactions between rare gas atoms, but the search of such representations for interactions between molecules is, of course, a more hazardous area. An essential element, in our opinion, is to represent such interactions by the means of contributions expressed in terms of local groups (bonds, atoms, lone pair orbitals) [ 30]. Indeed, this seems to be a good practical way for taking systematically into account the molecular shape, which are extremely variable (from this point of view, for example, the representation of the charge distribution of a molecule through a usual multipole expansion around a single center is useless, except when dealing with very large intermolecular distance : it is necessary to use a sum of local (atomic, bond...) charge distributions, each one of them being approximated by a small number of multipoles : we are thus led to the concept of a multi-centered multipole expansion [1, section V.A,B], [30]. Following the same line of thought, the polarization energy may be expressed as a sum of bond polarization terms (with the polarizing electric field created by the other molecule(s) being calculated e.g. at the middle of each bond), and the dispersion energy is expressed as a sum of bond-bond dispersion terms [42,43,44,30] [1, sections V.C,D,E]. Similarly, as concerns the short-range repulsion, an expression like (B.7), upon inserting localized (bond or lone pair) molecular orbitals, suggests the use of a sum of repulsion terms between bonds (and possibly lone pairs) [60,30],[1, section IV.E.2]. In all these cases, it may be tempting to reduce these formulae to sums of atom-atom terms (indeed, the atom-atom distances are anyway needed for evaluating the electrostatic interaction, and saving of computation time would be achieved if the computation of atom-bond and bond-bond distances could be avoided). We shall illustrate these considerations by summarizing two procedures (with different level of sophistication) developed in our laboratory :

(a) the simplest method (Caillet and Claverie [45], hereafter refered to as CC) is purely of the atom-atom type, i.e. the interaction energy is obtained as a sum of atom-atom terms only. This feature makes the computation time shorter than that needed by methods using a larger number of "centers of force" (e.g. in addition to the atoms, the middle of the bonds, or even the middles of all segments joining any pair of atoms), but this may be at the expense of the accuracy of the method, as we shall see below. We recall here only the main features of the method, refering to [45] for a complete description. The interaction energy is evaluated as a sum of four contributions :

(1) the electrostatic term is evaluated as the sum of interactions between atomic net charges (the problem here concerns the accuracy of the representation of the molecular charge distribution in terms of atomic charges only : see the method GCP below).

(2) the polarization term is evaluated as the sum of polarization energies of the atoms, each of them being endowed with a polarizability obtained by sharing bond polarizabilities, according to some weights deduced from the number of electrons that each atom brings to the bond considered. We use bond polarizabilities as our basic data because they are considered more transferable from one molecule to another, which makes it possible to use standard tabulated values [46, 47].

(3) the dispersion energy is evaluated as a sum of $1/R^6$ terms between atoms ($\lambda$ and $\nu$), and the short-range repulsion energy is evaluated as a sum of exponential terms exp $(-\alpha\, R_{\lambda\nu}/R^0_{\lambda\nu})$, where $R^0_{\lambda\nu}$ denotes some suitable "contact distance" characteristic of the atom pair $(\lambda,\nu)$. Otherwise stated, we have for representing these last two terms a sum of Buckingham type (6-exp) potentials between all pairs of atoms ($\lambda$ in one molecule, $\nu$ in the other). We followed here the procedure

proposed by Kitaigorodskii et al. 48,49,50], with some improvements (inclu ion in the repulsion term of factors taking into account the actual electronic population on the atoms, introduction of suitable modifications for short interatomic distances in order to get a satisfactory representation of hydrogen bonds [ 45]). Moreover, at very short distance, a suitable change of the dispersion, repulsion and polarization terms is introduced in order to avoid the spurious decrease to $-\infty$ of the 6-exp formula when the interatomic distance goes to zero [ 51, section 2-e ] (such a modification is necessary when performing energy minimization, or integrations infolving the intermolecular interaction potential, such as those occuring in statistical mechanical calculations). Applications of the CC method to crystals will be found in refs. [ 45, 51-56].

(b) the more refined method (Gresh, Claverie and Pullman [ 57], hereafter refered to as GCP) also evaluates the interaction energy as the sum of the same four contributions (electrostatic, polarization, dispersion and short-range repulsion), but the following improvements have been introduced :

(1) the charge distribution of each molecule is represented by a set of multipoles (charges, dipoles, quadrupoles) located at the atoms and at the middles of the segments defined by all pairs of atoms [ 1, sections V. A, B] [ 18, chap. III, section .B][ 30, section 5][ 58]. These multipoles may be obtained from accurate ab initio wave functions, and they reproduce correctly the essential features of the molecular electrostatic field [ 59]. The electrostatic term is now obtained as the interaction energy between the two sets of molecular multipoles. (In ref. [ 57], for the sake of simplicity, the dipoles and quadrupoles corresponding to non chemically bonded atoms were suppressed).

(2) as concerns the polarization energy, the electrostatic field of each molecule is now calculated from the corresponding multipoles, but, in addition, a larger set of "polarization centers" is used : not only the atoms, but also the middle of the bonds. Here also the corresponding polarizabilities are obtained by sharing the bond polarizabilities according to some effective number of electrons attributed to each center (bonding electrons, lone pairs). Moreover, in order to avoid the spurious divergences when some distance between centers $r_{ij}$ goes to zero, we use a modified distance $r'_{ij} = r_{ij} + 0.5\ D\ (W_i + W_j)$, where $W_i$ and $W_j$ are radii characteristic of the centers i and j, and D is some suitably fitted parameter (see [ 57] for details).

(3) the dispersion energy is evaluated as a sum of $1/(R'_{\lambda\nu})^6$ terms, the difference with respect to the CC method lying in the use of a "modified distance" $R'_{\lambda\nu} = R_{\lambda\nu} + 0.5\ F\ (W_\lambda + W_\nu)$ instead of the true interatomic distance $R_{\lambda\nu}$ (the situation is quite analogous to that for the polarization term : the use of $R'_{\lambda\nu}$ instead of R avoids the spurious divergence to $-\infty$ of the $1/(R_{\lambda\nu})^6$ formula when $R_{\lambda\nu} \to 0$).

(4) a more important difference occurs as concerns the short-range repulsion term : instead of a sum of atom-atom terms, a sum of bond-bond terms is used. Such a procedure is based upon the following theoretical argument [ 1, section IV.C.2, eq. (2.195)] , [ 18, chap. III, section 3.C.2 ][ 60] : the first-order exchange energy (which is the dominant contribution to the short-range repulsion term) could be approximately represented by the sum

$$C \sum_{i}^{(1)} \sum_{j}^{(2)} n_i^{occ} n_j^{occ} (S_{ij})^2 \qquad \text{(B.16)}$$

where $S_{ij}$ denotes the overlap integral between molecular orbitals i and j (belonging to molecule 1 and 2 respectively), $n_i^{occ}$ and $n_j^{occ}$ denote the occupation numbers and C is some suitable constant (a factor 1/R, suggested by the theory for the interactions between rare gas atoms, has been discarded for simplicity : it could be represented either through some suitably fitted value of C, or, more accurately, by using suitably modified pseudo-overlap integrals $S'_{ij}$ where the basic overlap integrals $S_{\alpha\beta}$ between atomic orbitals $\alpha$, $\beta$ would be replaced by $S'_{\alpha\beta} = S_{\alpha\beta}/ R_{\alpha\beta}$). Then, by taking into account the fact that the ground state wave function may be satisfactorily built from localized molecular orbitals (see e.g. [ 61], and further references in [ 1], section IV.E.2), the formula (B.16) immediately suggests the use of a sum of repulsion terms between bonds, corresponding to these localized orbitals (lone pair orbitals could also be considered). If these localized orbitals $\varphi_i = c_{i1} \chi_{\alpha 1} + c_{i2} \chi_{\alpha 2}$ and $\varphi_j = c_{j1} \chi_{\beta 1} + c_{j2} \chi_{\beta 2}$ are expressed in terms of atomic orbitals $\chi$, we get :

$$S_{ij} = c_{i1} c_{j1} S_{\alpha 1 \beta 1} + c_{i1} c_{j2} S_{\alpha 1 \beta 2} + c_{i2} c_{j1} S_{\alpha 2 \beta 1} + c_{i2} c_{j2} S_{\alpha 2 \beta 2} \qquad \text{(B.17)}$$

In order to build approximate formulae, this formula for $S_{ij}$ may be mimicked as a sum of four atom-atom exponential terms corresponding to the four atomic overlap integrals in (B.17), and this is just the procedure used in [ 57 ]. Some differences to be expected from the use of such bond-bond repulsion terms instead of the pure atom-atom terms were discussed perviously [ 1, section IV.E.2],[ 18, chap. III, section 3.C.2].

The results presented in [ 57 ] show that this refined method GCP seems able to reproduce correctly the essential features of interaction energy values obtained from SCF ab initio computations for systems involving a strong electrostatic interaction ($Na^+ ... H_2O$, $K^+ ... H_2O$, $NH_4^+ ... H_2O$ and $H_2O ... H_2O$). It is quite satisfactory that the hydrogen bond interaction may be reproduced without an ad hoc modification of parameters at short distance (such as was necessary in the CC method). Further improvements based upon theoretical considerations are still possible, noticeably introducing an explicit charge-transfer term (which would more accurately represent the short-range part of the complete induction energy), and, simultaneously, performing a suitable new parametrization of the short-range repulsion term (and possibly of the dispersion term) : it seems indeed likely that, at the present time, the short-range repulsion term implicitly includes at least some part of an attractive charge-transfer term, because it is smaller than the genuine 1st-order exchange term. It must be mentioned that the so-called "penetration" part of the exact Coulomb energy [62,63] plays a similar role, since it varies exponentially and is attractive in the medium- and long-range region. Thus, we conclude that numerous theoretical and practical improvements may be brought in the field of simplified formulae, but the results obtained so far already are rather satisfactory and encouraging.

## C. STUDY OF SOLVENT EFFECTS

### 1. Theoretical fremawork

The treatment of solvent effects (and, more generally, of problems involving liquids) presents the following well-known basic difficulty : it is no more possible to restrict one's attention to two molecules only (as is the case in dilute gases), since each molecule is directly surrounded by many others (first, second,...solvation shells), and on the other hand it is not possible to assume from the outset that the molecules lie at well-fixed positions (as is the case in crystals- : some non-trivial statistical treatment (or at least some minimization of the energy with respect to molecular configurations) has to be performed : even if some more or less structured solvation shell exists, this has to be proved first by some adequate calculation.

Typical solvent effects concern the relative stability of the conformations of a solute molecule, or the formation of intermolecular complexes between two solute molecules. In all cases, we have two species denoted I and II : two conformations in the first case, the species "separated solute molecules" and "complex surrounded by the solvent" in the second case. The energy difference $\Delta E_{vac} = E_{II} - E_I$ between the species I and II in vacuum is supposed to be known, and we want to evaluate the corresponding differences between the thermodynamic functions in solution : $\Delta E_{sol}$, $\Delta F_{sol}$ (free energy) or rather $\Delta G_{sol}$ (free enthalpy). For that purpose, we define the solvation thermodynamic quantities ($\Delta E_{sol}$, $\Delta F_{sol}$, $\Delta G_{sol}$) for each species, i.e. the variation of E, F, G upon transfer of the species from vacuum to the solvent, and we then obtain :

$$\Delta E_{sol} = \Delta E_{vac} + \left[\Delta E_{solv}(II) - \Delta E_{solv}(I)\right] \qquad (C.1)$$

$$\Delta F_{sol} = \Delta F_{vac} + \left[\Delta F_{solv}(II) - \Delta E_{solv}(I)\right] \qquad (C.2)$$

$$\Delta G_{sol} = \Delta G_{vac} + \left[\Delta G_{solv}(II) - \Delta G_{solv}(I)\right] \qquad (C.3)$$

The bracketed quantities such as $\Delta E_{solv}(II) - \Delta E_{solv}(I)$ precisely represent the solvent effect.

In $\Delta F_{vac}$ and $\Delta G_{vac}$, the entropic part essentially corresponds to the loss of translational and rotational degrees of freedom and their replacement by vibrational degrees of freedom in the case of complex formation. This part may be approximately the same in vacuum and in the solvent, anyway we shall be concerned here essentially with the evaluation of the solvation quantities, and more precisely the quantities associated with the insertion of the solute species at some fixed position in the solvent, according to the analysis by Ben Naim [64] : in otherwords, we discard the so-called liberation free energy, which is associated with translation (and exchange) of the solute molecules, since it is not at all specific of the intermolecular interactions, and could be easily evaluated if needed [64]. Thus, our $\Delta G_{solv}$ for 1 mole corresponds to the chemical potential $\tilde{\mu}$ of Ben Naim [64]. By the way, we shall suppose in the sequel that our solution is very dilute, but this restriction is not essential : it entails the simplification that, when considering the solute-solvent interactions, the medium surrounding the solute molecule may be endowed with the properties (dielectric constant,...) of the pure solvent.

## 2. Statistical mechanical treatment of solvation : charging parameter method, cavitation and solute-solvent interaction.

How to evaluate the solvation thermodynamic quantities ? Ben Naim [64] expresses them directly as statistical averages (denoted $< >_0$), which involve the distribution function of the solvent (with the additional solute molecule fixed, namely (eq. (3.12) of [64]) :

$$\tilde{\mu}_A = -RT \text{ Log } < \exp - B_A(\vec{q}_0)/RT >_0 \qquad \text{(C.4)}$$

where $\tilde{\mu}_A = \Delta G_{solv}$ (A) for 1 mole of solute A, and $B_A(\vec{q}_0)$ denotes the interaction energy between the solvent and the solute fixed at some configuration $\vec{q}_0$ ($\vec{q}$ stands for $(\vec{r},\vec{\omega})$, where $\vec{r}$ denotes the position and $\vec{\omega}$ the orientation).

The evaluation of the statistical average $< >_0$ could conceivably be done through the Monte-Carlo method.

For our own purpose, which involves the development of approximate (continuum) models, we use the "charging parameter" method [65a -68] [69, appendix C] [70, appendix B] [71, appendix ]. We consider two thermodynamic states (0) and (1) characterized by the total intermolecular interaction energy functions $U_0(\vec{Q})$ and $U_1(\vec{Q})$ respectively : for example, $U_0$ corresponds to dispersion and short-range repulsion only, and $U_1$ corresponds to the complete interaction energy with electrostatic and polarization terms added (this case essentially corresponds to the Debye-Hückel treatment of ionic solutions [65b, 67]; or $U_0$ corresponds to the complete interaction energy of the solvent only, while $U_1$ corresponds to the energy of the solvent plus one molecule of the solute in some fixed configuration : $U_1 = U_0 + B_A(\vec{q}_0)$. Then, a transformation from state 0 to state 1 is defined by introducing an interaction energy function $U(\vec{Q},\xi)$ depending on the so-called "charging parameter" $\xi$, in such a way that :

$$U(\vec{Q},0) = U_0(\vec{Q}) \quad \text{and} \quad U(\vec{Q},1) = U_1(\vec{Q}) \qquad (C.5)$$

A trivial procedure consists in defining :

$$U(\vec{Q},\xi) = U_0(\vec{Q},\xi) + \xi\Delta U(\vec{Q}) \qquad (C.6)$$

where $\Delta U(\vec{Q}) = U_1(\vec{Q}) - U_0(\vec{Q})$. Then, by using the expression of the free energy $F(\xi)$ (corresponding to the canonical ensemble, see e.g. [ 66] ):

$$F(\xi) = -RT \operatorname{Log} \int \exp\left[- U(\vec{Q},\xi)/RT\right] d\vec{Q} \qquad (C.7)$$

we easily get (see e.g. [ 69], appendix C) :

$$\frac{\partial F}{\partial \xi} = \left\langle \frac{\partial U}{\partial \xi} \right\rangle_\xi \text{ hence } \Delta F = F_1 - F_0 = \int_0^1 \left\langle \frac{\partial U}{\partial \xi} \right\rangle_\xi^C d\xi \qquad (C.8)$$

where $\langle\ \rangle_\xi^C$ denotes the canonical average corresponding to the interaction energy $U(\vec{Q},\xi)$, i.e. to the weight $\exp\left[-U(\vec{Q},\xi)/RT\right]$. Note that, for the special choice (C.6) of $U(\vec{Q},\xi)$, we get $\frac{\partial U}{\partial \xi} = \Delta U(\vec{Q})$, so that the general formula (C.8) becomes

$$\Delta F = \int_0^1 \langle \Delta U \rangle_\xi^C \, d\xi \qquad (C.9)$$

From $\Delta F$, all other thermodynamic functions may be obtained, e.g. (Gibbs-Helmholtz relation) :

$$\Delta E = \Delta F - T \frac{\partial(\Delta F)}{\partial T} \quad \text{and} \quad \Delta S = - \frac{\partial(\Delta F)}{\partial T} \qquad (C.10)$$

The free enthalpy $\Delta G$ may be obtained as $\Delta F + \Delta(PV)$, or it may be expressed by a formula analogous to (C.8) through the use of the isothermal-isobaric ensemble (T,P fixed) [66a, 64] instead of the canonical one (T,V fixed) : thus $\langle\ \rangle_\xi^C$ would simply be replaced by $\langle\ \rangle_\xi^{TP}$ . In actual fact, it may be argued that, when applying this method to the continuum model (see below), we shall rather be in the second case, because the quantities of experimental origin used in this model (dielectric constant and its derivative with respect to temperature, thermal expansion coefficient...) are usually measured for fixed T and P rather than fixed T and V. Anyway, for comparing two species I and II in the solution, the difference $\Delta(PV)$ between $\Delta G$ and $\Delta F$ is usually negligible in practice.

We shall now express the thermodynamic functions of solvation through a (virtual) process involving two main steps :

(a) cavitation step : initial state : the pure solvent (solute-solvent interaction completely "switched off"); final state : the solvent surrounding a hard solute molecule, i.e. a molecule interacting via a hard-core type potential (e.g. a sum of hard-sphere potentials corresponding to the various atomic Van der Waals spheres, endowed with suitable radii). This step therefore corresponds to the "switching-on" of this hard-core solute-solvent interaction $U_{hc}(\vec{q}_0,\vec{Q})$ ($\vec{q}_0$ : configuration of the solute; $\vec{Q}$ : configuration of all solvent molecules), and the final state corresponds to the creation in the solvent of a cavity with the size and shape of the solute molecule, hence the name "cavitation step". Note that the simple choice (C.6) for $U(\vec{Q},\xi)$ cannot be used here, since $U_1$ is infinite inside the hard-core. The reason for introducing this step is that it involves the physical properties of the solvent only : the solute appears only through its geometrical properties.

(b) The solute-solvent interaction step corresponds to the "switching-on" of the genuine solute-solvent interaction energy $U_{s-sv}(\vec{q}_0,\vec{Q})$, or, more precisely, of the difference $U_{s-sv}-U_{hc}$ with the previously switched-on hard-core itneraction. In practice, it is convenient to subdivide this step into two substeps (see [ 71], appendix, part 2 : Güntelberg-Müller "charging" process) :

(b1) "switching-on" of the dispersion-repulsion part $U^{dis-rep}_{s-sv}(\vec{q}_0,\vec{Q})$(minus the hard-core interaction $U_{hc}$). According to the analysis presented in [ 71] and [ 69, appendix C], if the change of the distribution function (of the solvent molecules) during this step is neglected, we have $\Delta F = \Delta E$, and since the contribution of $U_{hc}$ to the average energy is zero (in the initial state), we would get approximately for this step (see [ 71], eq. A.10) :

$$\Delta E_1 \simeq \Delta F_1 \simeq < U^{dis-rep}_{s-sv} >_{hc} \tag{C.11}$$

where the average is taken with the (assumed unchanged) distribution function of the solvent molecules corresponding to the hard-core solute.

(b2) "switching on" of the electrostatic (and polarization) part $U^{el(p)}_{s-sv}(\vec{q}_0,\vec{Q})$ of the solute-solvent interaction energy). Here, the solvent reorganization which is associated with the change of the distribution function of the solvent molecules, can by no means be neglected when the "charging parameter" $\xi$ varies from 0 to 1, and consequently $\Delta E_2 \neq \Delta F_2$. Note that, here, the simple choice (C.6) can be used, with $\Delta U = U^{el(p)}_{s-sv}$, and accordingly $\Delta F_2$ will be given by (C.9)

$$\Delta F_2 = \int_0^1 < U^{el(p)}_{s-sv} >_{\xi} d\xi \tag{C.12}$$

It must be strongly emphasized that $\Delta F_2$ really is the full free energy change associated with the "switching-on" of $U^{el(p)}_{s-sv}$, including the contribution due to the solvent reorganization, despite the fact that the solvent-solvent interaction energy does not explicitly appears in (C.12) (this result was also obtained recently by Blaive [ 72 ] through thermodynamical arguments in the case of dielectric media). The contribution of the solvent reorganization is actually taken into account through the change of the distribution function of the solvent molecules, a change which is involved in the ξ-dependent average $< >_\xi$. This point was not clearly appreciated in the section 2.b of the Appendix in [71] : all calculations are correct, but the comments between eqs. (A.11) and (A.12) are not and must be suppressed. Some confusion may have arisen from the fact that the full energy difference $\Delta E_2 = \Delta F_2 - T\,\partial(\Delta F_2)/\partial T$ may be decomposed into two parts : the average interaction energy between the solute and the reorganized solvent, which is obviously given by $< U^{el(p)}_{s-sv} >_1$, and the reorganization energy of the solvent $\Delta E_{reorg}$ (sv) which is the change of the average solvent-solvent interaction energy $< U_{sv-sv} >$ between the states $\xi = 0$ and $\xi = 1$ : thus, we see that this energy may be obtained as the difference $\Delta E_2 - < U^{el(p)}_{s-sv} >_1$.

When applied to a pure liquid, this analysis provides us with an expression for the cavitation quantities $\Delta E_{cav}$, $\Delta F_{cav}$, $\Delta G_{cav}$. Indeed, the vaporization quantities are known :
at liquid-vapor equilibrium :

$$\Delta G_{vap} = 0, \quad \Delta F_{vap} = -\Delta(PV) = -P\Delta V \simeq -RT \text{ and}$$
$$\Delta E_{vap} = -\frac{1}{2} < U^{total}_{s-sv} > \tag{C.13}$$

where s now denotes just one of the molecules of the pure liquid, considered as a solute molecule : indeed, when dealing with pairwise interactions :

$$U^{total} = \sum_{i<j} U_{ij} = \frac{1}{2} \sum_{i \neq j} U_{ij} = \frac{N}{2} \sum_{j \neq i} U_{ij} = \frac{N}{2} U^{total}_{s-sv}$$

hence the vaporization energy per molecule $\Delta E_{vap} = - < U^{total} >/N = -\frac{1}{2} < U^{total}_{s-sv} >$ (see [1] section VI.E, and [45] section 2.B.a for related arguments). On the other hand, $\Delta E_{vap}$ is just the opposite of $\Delta E_{solv}$ the solvation energy of a molecule in its own liquid, and similarly for ΔF and ΔG : hence, by using the expressions obtained above for $\Delta E_{solv}$, $\Delta F_{solv}$, or $\Delta G_{solv}$, in terms of a cavitation part and two solute-solvent interaction parts, we shall get equations involving the cavitation quantities :

$$-\Delta E_{vap} = \Delta E_{cav} + \Delta E_1 + \Delta E_2 \tag{C.14}$$

gives, according to (C.13)

$$\Delta E_{cav} \simeq \frac{1}{2} < U^{total}_{s-sv} > - (\Delta E_1 + \Delta E_2) \tag{C.15}$$

with $\Delta E_1$ given by (C.11) and $\Delta E_2$ by (C.12) and (C.10), and similarly :

$$- \Delta F_{vap} = \Delta F_{cav} + \Delta F_1 + \Delta F_2 \quad (C.16)$$

gives

$$\Delta F_{cav} \simeq \Delta RT - (\Delta F_1 + \Delta F_2) \quad (.17)$$

with $\Delta F_1$ given by (C.11) and $\Delta F_2$ by (C.12) (for $\Delta G_{cav}$, we would suppress RT and replace the canonical averages by (T,P) averages in (C.11) and (C.12)). The results given by such formulas for the cavitation thermodynamic quantities could be checked against those provided by different approaches : Monte-Carlo computations [ 73], scaled particle theory [ 74].

### 3. The "discrete", "continuum" and mixed (discrete-continuum) models

The theoretical developments of the previous section were essentially exact (except for some approximate formulae such as (C.11)). But in order to treat practical problems involving large molecules, the use of accurate computational methods (Monte Carlo, Molecular Dynamics) for evaluating the various statistical averages may become too costly, hence the incentive for developing approximate methods. We were led to propose three such methods in previous works [ 69-71,75 ] , and we shall briefly recall here their main features in the theoretical framework of the previous section :

(a) the "discrete" model : several solvent molecules are placed around the solute, but usually not numerous enough for really simulating the whole solvent (example : in [ 71], 4 to 57 water molecules were put around $NH_4^+$). Moreover, up to now, energy minimizations only (and not statistical mechanical averaging) were performed. Thus, the entropy-dependent functions ($\Delta F$ or $\Delta G$) are not obtained, and the model seems therefore suited for cases where the energy $\Delta E$ plays the essential role (study of the structure of the first solvation shells, noticeably for ionic groups in water). Examples of applications : $NH_4^+$ and $CH_3OH$ in water [ 71], Acetylcholine cation in water [ 75], DiMethylPhosphate anion ($DMP^-$) in water [ 76], Purine and bimolecular purine complex in water [ 77].

(b) the "continuum" model : the basic idea is to avoid the complexities of the exact statistical mechanical treatment (discrete summations over a large number of molecules and statistical averages) by representing the solvent as a continuous medium endowed with suitable properties (polarizability per unit volume, dielectric constant, thermal expansion coefficient...), while the solute is still treated as a genuine molecule (involving electric multipoles, polarizable centers, dispersion and repulsion force centers), with a suitable Van der Waals volume (outside which lies the continuous medium). In early works, simple-shaped cavities (spherical, ellipsoidal) only were considered, but we developed [ 69,75,70,71] the theory for arbitrary cavities (in practice, we use the union of atomic Van der Waals spheres as our molecular cavity). Discrete summations may be replaced by integrations over the continuous medium, but there is still the problem of simulating in an appropriate way the statistical averaging. For the dispersion and repulsion terms, the distribution function of the solvent molecules (and atoms) is considered as uniform [69] [71, appendix, part 2.a], while, for the electrostatic (and polarization) terms, the solvent is considered as a dielectric medium (polarized by the multipoles moments of the solute molecule), and the problem is thus reduced to the evaluation of the reaction field (cf.[68]) through classical methods of electrostatics [75, 70]. An accurate solution is still somewhat cumbersome for large molecules, because the basis of spherical harmonics (used for expressing the electrostatic potential outside the cavity) has a dimension proportional to the number of the atoms of the solute molecule [75], thus more approximate methods (reducing the problem to a single integration over the dielectric continuum, or rather the surface of the cavity) are being investigated [75, section VI; 77 , 78]. The evaluation of the cavitation term is one of the main difficulties in continuum models at the present time : approaches based upon the macroscopic surface tension, even when including some microscopic correction term [79, 80] do not seem satisfactory ($\Delta E_{cav}$ seems too large), and we prefered to use, in our recent works [71, 76, 77, 81] Pierotti's formulae [74] based upon scaled-particle theory. As indicated at the end of the previous subsection 2, it appears now possible to evaluate the cavitation term for a molecule in its own liquid, entirely in the framework of the continuum model, by evaluating (according to this model) all the necessary statistical averages, and this should provide a check of the consistency of the model with the other approaches to cavitation. For examples of applications of the continuum model, see [69-71, 75-77,81]. In [70], the vaporization energy of pure polar liquids was considered, but through a direct evaluation (eq. (C.13)), and not through the cavitation and solute-solvent steps : consequently $\Delta E_{vap} = -1/2 \langle U^{dis-rep}_{s-sv} \rangle - \frac{1}{2} \langle U^{el(p)}_{s-sv} \rangle$ should have been used instead of $-\frac{1}{2} \langle U^{dis-rep}_{s-sv} \rangle - \Delta E_2$ (where $\Delta E_2$ is obtained through (C.10) and (C.12)). When $\Delta E_2$ is replaced by $\frac{1}{2} \langle U^{el(p)}_{s-sv} \rangle$, more satisfactory results are actually obtained (thus, for non-associating liquids, the empirical factor f may be taken equal to 1, instead of the value 0.33 which was found necessary [70] when using $\Delta E_2$).

(c) <u>The mixed ("discrete-continuum") model</u> : Here, in order to better represent specific interactions between the solute and the closest solvent molecules, a limited number of these is explicitly treated as discrete (e.g. those of the first solvation shell), while the rest of the solvent is represented by a continuous medium. The price to pay for this more faithful representation of the solvent is, of course, the need to perform energy minimizations or statistical averages concerning those solvent molecules which are explicitly treated as discrete. Nevertheless, their small number should make the task much easier than it would be for a completely discrete treatment, and the mixed model therefore appears as a satisfactory compromise between the opposite qualities and defects of the two "extreme" models (purely discrete and purely continuous). For some preliminary applications, see [71,76,77, 81]. Following the same line of thought, we are also trying to improve the continuum model itself by allowing for a better representation of the closest solute-solvent interactions, noticeably as concerns the problem of the variation of the equilibrium distance between atoms according to the relative weight of non polar and polar (e.g. hydrogen bond) interaction terms.

Acknowledgements. I am grateful to B. Blaive for discussions about his thesis, which helped me to clear up several points concerning the "charging parameter" method.

## REFERENCES

[1] Claverie, P.: 1978, in "Intermolecular Interactions : from Diatomics to Biopolymers", ed. by B. Pullman, Wiley, New York; Chap. 2, pp. 69-305.
[2] Ahlrichs, R. : 1976, Theoret. Chim. Acta, 41, pp. 7-15.
[3] Claverie, P.:1971, Intern. J. Quant. Chem. 5, pp.273-296.
[4] Chalasinski, G., Jeziorski, B. and Szalewicz, K.:1977, Intern. J. Quant. Chem. 11, pp.247-257.
[5] Jeziorski, B., Szalewicz, K. and Jaszunski, M.:1979, Chem. Phys. Letters, 61, pp.391-395.
[6] Jeziorski, B.: 1979, "What is the optimal symmetry-adapted perturbation theory ? Adequacy of the weak symmetry forcing". Communication presented at the "Sanibel Quantum Chemistry Symposium".(March 1979).
[7] Jeziorski, B. and Kolos, W.:1977, Intern. J. Quant. Chem. 12, suppl. 1, pp.91-117.
[8] Jeziorski, B., Szalewicz, K. and Chalasinski, G.:1978, Intern. J. Quant. Chem. 14, pp.271-287.
[9] Kutzelnigg, W.:1977, "Explanation of the failure of some versions of exchange perturbation theory", preprint.
[10] Kutzelnigg, W.:1978, Chem. Phys. 28, pp.293-298.
[11] Kutzelnigg, W.:1978, Intern. J. Quant. Chem. 14, pp.101-120.
[12] Kutzelnigg, W.:1978, "The "primitive" wave function in the theory of intermolecular interactions", preprint (a condensed version will be published).
[13] Certain, P.R., Hirschfelder, J.O. and Epstein, S.T.: 1969, Chem. Phys. Letters, 4, pp.401-404.
[14] Certain, P.R., Hirschfelder, J.O., Kolos, W. and Wolniewicz, L. : 1968, J. Chem. Phys. 49, p.24-34.
[15] Certain, P.R.: 1968, J. Chem. Phys. 49, p.35-38.
[16] Basilevsky, M.V. and Berenfeld, M.M.: 1972, Intern. J. Quant. Chem. 6, p.23-45.
[17] Chipman, D.M.: 1972, "Perturbation theories for the calculation of molecular interaction energies", Ph.D. Thesis, University of Wisconsin, Madison.
[18] Claverie, P.:1973, "Contribution à l'Etude des Interactions Molécu-

laires", Thèse de Doctorat d'Etat, Paris (registration number at the C.N.R.S. library : A.O. 8214).
[19] Chipman, D.M. and Hirschfelder, J.O.:1973, J. Chem. Phys. 59, pp. 2838-2857.
[20] Snook, I.K. and Spurling, T.H.:1975, J. Chem. Soc. Faraday Trans. II, 71, pp.852-861.
[21] Chalasinski, G. and Jeziorski, B.:1976, Molec. Phys. 32, pp.81-91.
[22] Chalasinski, G., Jeziorski, B., Andzelm, J. and Szalewicz, K.:1977, Molec. Phys. 33, pp.971-977.
[23] Szalewicz, K. and Jeziorski, B.:1979, Molec. Phys. 38, pp.191-208.
[24] Jeziorski, B. and Van Hemert, M.:1976, Molec. Phys. 31, pp.713-719.
[25] Murrell, J.N. and Shaw, G.:1967, Molec. Phys. 12, pp.475-480 (erratum : Molec. Phys. 15, p.325 (1968)).
[26] Whitton, W.N. and Byers Brown, W.:1976, Intern. J. Quant. Chem. 10, pp.71-86.
[27] Murrell, J.N. and Shaw, G.:1967, J. Chem. Phys. 46, pp.1768-1772.
[28] Musher, J.I. and Amos, A.T.:1967, Phys. Rev. 164, pp.31-43.
[29] Messiah, A.:1959 and 1960, "Mécanique Quantique", Dunod, Paris (vol. I:1959; Vol.II:1960). English translation : "Quantum Mechanics", North Holland, Amsterdam (1961).
[30] Claverie, P.:1976, in "Localization and Delocalization in Quantum Chemistry", vol. II, ed. by O. Chalvet, R. Daudel, S. Diner and J.P. Malrieu, Reidel, Dordrecht, pp.127-152.
[31] Longuet-Higgins, H.C.:1956, Proc. Roy. Soc. (London) A235, pp.537-543.
[32] Kauzmann, W.:1957, "Quantum Chemistry", Academic Press, New York. Chap. 16, section A.c, pp.645-646.
[33] Dalgarno, A. and Davison, W.D.:1966, in "Advances in Atomic and Molecular Physics", vol. 2 (ed. by D.R. Bates and I. Estermann), Academic Press, New York, pp.1-32.
[34] Murrell, J.N., Randic, M. and Williams, D.R.:1965, Proc. Roy. Soc. (London) A284, pp.566-581.
[35] Cohen, J.S. and Pack, R.T.:1974, J. Chem. Phys. 61, pp.2372-2382.
[36] Murrell, J.N. and Varandas, A.J.C.:1975, Molec. Phys. 30, pp.223-236.
[37] Mavroyannis, C. and Stephen, R.J.:1962, Molec. Phys. 5, pp.629-638.
[38] Salem, L.:1960, Molec. Phys. 3, pp.441-452.
[39] Tang, K.T. and Karplus, M.:1968, Phys. Rev. 171, pp.70-74.
[40] Dalgarno, A.:1967, in "Intermolecular Forces" (ed. by J.O. Hirschfelder), Advances in Chemical Physics, vol.12, pp.143-166 (Wiley-Interscience, New York).
[41] Kim, Y.S. and Gordon, R.G.:1961, J. Chem. Phys. 61, pp.1-16.
[42] Rein, R., Claverie, P. and Pollak, M.:1968, Intern. J. Quant. Chem. 2, pp.129-144.
[43] Claverie, P. and Rein, R.:1969, Intern. J. Quant. Chem. 3, pp.537-551.
[44] Amos, A.T. and Crispin, R.J.:19 (a) Molec. Phys. 31, pp.159-176; (b) in "Theoretical Chemistry, Advances and Perspectives", vol.2 (H. Eyring and D. Henderson eds.), Academic Press, New York, pp.1-66.

[45] Caillet, J. and Claverie, P.:1975, Acta Cryst. A31, pp.448-461.
[46] Le Fèvre, R.J.W.:1965, in "Advances in Physical Organic Chemistry" vol. 3, ed. by V. Gold, Academic Press, New York : pp.1-90.
[47] Denbigh, K.G.:1940, Trans. Faraday Soc. 36, pp.936-948.
[48] Kitaigorodskii, A.I.:1961, Tetrahedron, 14, pp.230-236.
[49] Kitaigorodskii, A.I. and Mirskaya, K.V.: Sov. Phys.- Crystallogr. 9, 137-142.
[50] Kitaigorodskii, A.I., Mirskaya, K.V. and Nauchitel, V.V.:1970, Sov. Phys. Crystallogr. 14, pp.769-771.
[51] Caillet, J., Claverie, P. and Pullman, B.:1967, Acta Cryst. B32, pp.2740-2745.
[52] Caillet, J. and Claverie, P.:1974, Biopolymers, 13, pp.601-614.
[53] Caillet, J., Claverie, P. and Pullman, B.:1977, Acta Cryst. A33, pp.885-889.
[54] Caillet, J., Claverie, P. and Pullman, B.:1978, Theoret. Chim. Acta 47, pp.17-26.
[55] Caillet, J., Claverie, P. and Pullman, B.:1978, Acta Cryst. B34, pp.3266-3272.
[56] Caillet, J. and Claverie, P.:1980, "On the polymporphism of barbituric acid derivatives", Acta Cryst. B, in press.
[57] Gresh, N., Claverie, P. and Pullman, A.:1979, Intern. J. Quant. Chem. Quant. Chem. Symp. 13, pp.243-253.
[58] (a) Dreyfus, M.: 1970,"Etude non-empirique de la liaison hydrogène dans les dimères de la formamide", thèse de 3è cycle, (Université Paris VI), chap. II (b) Port, G.N.J. and Pullman, A.: 1973, FEBS Letters, 31, pp.70-74.
[59] Goldblum, A., Perahia, D. and Pullman, A.:1979, Int. J. Quant. Chem. 15, p.121-129.
[60] Salem, L.: 1961, Proc. Roy. Soc. (London), A264, pp.379-391.
[61] Chalvet, O., Daudel, R., Diner, S. and Malrieu, J.P. (editors) : 1975, "Localization and Delocalization in Quantum Chemistry", Reidel, Dordrecht. See especially the papers by P. Millié, B. Levy and G. Berthier, pp.59-97, and by J.P. Daudey, J.P. Malrieu and O. Rojas, pp.155-205.
[62] Van Duijneveldt, F.B. and Murrell, J.N.:1967, J. Chem. Phys. 46, pp.1759-1767.
[63] Van Duijneveldt, F.B.:1968, J. Chem. Phys. 49, pp.1424-1429.
[64] Ben-Naim, A.:1978, J. Phys. Chem. 82, pp.792-803.
[65] Fowler, R.H.:1936 "Statistical Mechanics", Cambridge University Press : (a) section 8.8 pp.267-269; (b) section 8.81, pp.269-274; (c) sections 13.6, 13.61,13.62, 13.7 pp.541-552.
[66] Hill, T.L.:1956 "Statistical Mechanics" McGraw-Hill, New York; (a) chap. 3, sections 14-15, pp.59-80, (b) chap. 6, sections 30-31 pp.191-195.
[67] Eyring, H., Henderson, D., Stover, B.J. and Eyring, E.M.:1964, "Statistical Mechanics and Dynamics", Wiley, New York:chap. 14.
[68] Linder, B.:1967, Adv. Chem. Phys. 12, pp.225-282: (a) p.232; (b) pp.261-264.
[69] Huron, M.J. and Claverie, P.:1972, J. Phys. Chem. 76, pp.2123-2133.
[70] Huron, M.J. and Claverie, P.:1974, J. Phys. Chem. 78, pp.1862-1867.

[71] Claverie, P., Daudey, J.P., Langlet, J., Pullman, B., Piazzola, D., and Huron, M.J.:1978, J. Phys. Chem., 82, pp.405-418.
[72] Blaive, B.:1980, "Contribution à la Théorie de la Polarization Electrique et à l'Etude des Interactions Electrostatiques des Molécules d'un liquide", Thèse de Doctorat d'Etat, Université d'Aix-Marseille III, Chap. X et XIV.
[73] Dashevsky, V.G. and Sarkisov, G.N.:1974, Mol. Phys. 27, pp.1271-1290.
[74] Pierotti, R.A.:1976, Chem. Rev. 76, pp.717-726.
[75] Huron, M.J. and Claverie, P.:1974, J. Phys. Chem. 78, pp.1853-1861.
[76] Langlet, J., Claverie, P., Pullman, B., Piazzola, D. and Daudey, J.P.:1977, Theoret. Chim. Acta, 46, pp.105-116.
[77] Langlet, J., Giessner-Prettre, C., Pullman, B. and Claverie, P.: 1980, "Purine-Water Interactions in Base Stacking", Intern. J. Quant. Chem., Proceedings of the 3rd International Quantum Chemistry Congress (Kyoto, 1979), in press.
[78] Miertus, S., Scrocco, E. and Tomasi, J.: "Electrostatic Interaction of one Molecule with continuum", Chem. Phys. (1981).
[79] Sinanoglu, O.:1968, in "Molecular Associations in Biology", ed. by B. Pullman, Academic Press, New York, pp.427-445.
[80] Beveridge, D.L., Kelly, M.M. and Radna, R.J.:1974, J. Amer. Chem. Soc. 96, pp.3769-3778.
[81] Langlet, J., Claverie, P., Pullman, B. and Piazzola, D.:1979, Intern. J. Quant. Chem. Quant. Biol. Symp. 6, pp.409-437.

## INDEX OF SUBJECTS

GPSR Compliance

*The European Union's (EU) General Product Safety Regulation (GPSR) is a set of rules that requires consumer products to be safe and our obligations to ensure this.*

*If you have any concerns about our products, you can contact us on ProductSafety@springernature.com*

In case Publisher is established outside the EU, the EU authorized representative is:

Springer Nature Customer Service Center GmbH
Europaplatz 3
69115 Heidelberg, Germany

**Batch number: 10584448**

Printed by Printforce, the Netherlands